An ABC Resource Survey [illegible]
for
Environmentally Significant Areas
with
Special Reference to Biotic Surveys
in Canada's North

Department of Geography Publication Series No. 24
University of Waterloo

An ABC Resource Survey Method for Environmentally Significant Areas with Special Reference to Biotic Surveys in Canada's North

by Jamie D. Bastedo

All artwork prepared in the Faculty of Environmental Studies Cartographic Centre.

Canadian Cataloguing in Publication Data

Bastedo, Jamie, 1955-
An ABC resource survey method for environmentally significant areas with special reference to biotic surveys in Canada's North

(Department of Geography publication series ; no. 24)
Bibliography: p.
ISBN 0-921083-20-3

1. Ecological surveys - Yukon Territory. 2. Ecology - Yukon Territory. I. University of Waterloo. Dept. of Geography. II. Title. III. Series.

QH541.15.S95B37 1986 574.5'028 C86-093329-6

Published by the Department of Geography

Abstract

To provide baseline information pertinent to broad scale (1:125,000) land use planning for Environmentally Significant Areas (ESA) in Canada's Yukon Territory, a resource survey approach was developed which involves the independent analysis and subsequent integration of abiotic, biotic and cultural (ABC) resource data. While intended for application to ESAs in the Yukon, the general approach may be applied, in whole or in part, to parks and reserves elsewhere.

The ABC approach was developed primarily from two bases: (a) a planning philosophy common to parks and reserves generally and to the Yukon's ESAs in particular; and (b) an appraisal of existing resource survey types in the context of six pre-defined criteria: economy, flexibility, replicability, ecological validity, communicability and applicability.

Results of preliminary research indicated that a hybrid approach, amalgamating many of the methodological strengths of existing survey types, was most appropriate for ESAs. Important considerations incorporated into the approach's design include: selection of data variables which reflect both structural and functional aspects of the environment; formulation of distinct but complementary classification schemes for categorizing and mapping abiotic, biotic and cultural land units; adoption of data acquisition procedures which capture the range of ecological diversity within ESAs; development of data analysis procedures which allow for the differentiation of land units based on their relative importance in supporting ecosystem self-maintenance; and organization of resource information and related recommendations into a form which is meaningful to both lay and technical audiences. The main elements of the proposed ABC approach are presented within the context of a conceptual model; specific procedures for the biotic part of this approach are presented subsequently.

According to the biotic procedures described, raw (descriptive) data on vegetation and wildlife are collected and analyzed by relying primarily on LANDSAT satellite imagery, field investigations and a review of previous research. Raw data are then translated into six interpretive indices related to biotic resource values. These indices include community diversity, uniqueness, recoverability, faunal diversity, faunal dependence and fire susceptibility. Then, through a systematic but flexible evaluation process, indices' scores (excluding fire susceptibility) for each biotic land unit are combined in such a way as to differentiate units with respect to their relative biotic significance - high, medium or low. Each land unit, or group of land units, of high biotic significance is discussed with respect to: (1) their specific biotic resource values; (2) the degree to which human activities (past, present and future) may influence these values; (3) immediate management needs for conserving these values; and (4) research needs and opportunities presented by these values.

In appraising the proposed ABC resource survey approach, the author concludes that by incorporating a process of inductive resource analysis into its design, the approach is more ecologically valid, and hence, more applicable to land use planning in ESAs and related reserves than other approaches which rely more upon current perceptions for recognizing ecosystems. Results of a complete ABC resource survey as here described could reveal the degree of interrelatedness both within and between abiotic, biotic and cultural components of the environment, thus aiding in the recognition of ecological linkages among land units and suggesting boundaries for appropriately-sized planning or management areas. Details on other possible applications of survey results such as land use control mechanisms and institutional arrangements for the establishment and management of ESAs must await the fruits of ongoing research being undertaken by the ESA research team at the University of Waterloo, Waterloo, Ontario, Canada.

Preface

John B. Theberge
School of Urban and Regional Planning
University of Waterloo
Waterloo, Ontario

The protection of the best examples of Canada's rich diversity of ecosystems and its wildland and wildlife heritage requires social values that embrace things "natural, wild and free". It also requires a science which identifies the lands best suited for these purposes. This monograph advances the latter requirement by describing a method of resource survey specifically designed for environmentally significant areas worth protecting as parks or related reserves or through other institutional arrangements.

Ecologists, naturalists and others who understand the importance of natural systems often look with dismay at the slow progress in ecosystem preservation compared with development, despite many governmental guidelines calling for "balanced development". The growth of our national, provincial and territorial parks, wildlife refuges and ecological reserves is painfully slow. Increasingly, people are becoming concerned by the considerable evidence that we are squandering our wildlands heritage. This heritage is one of the fundamental characteristics of Canada and the Canadian identity. Historically, our piecemeal approach to establishment of parks and reserves has not only been slow, but has resulted in considerable conflict with other land users. There must be better ways to identify and allocate land uses other than pitting the uncoordinated activities of a host of government agencies, each with some piece of the conservation mandate, against the opportunistic adventurism of development interests.

The Yukon Environmentally Significant Areas Research, directed by J.G. Nelson and myself, was developed specifically to respond to a major recommendation made by the Parks and Reserves Working Group of the Canadian Arctic Resources Committee (CARC) National Workshop held in Edmonton, Alberta, in February, 1978. The recommendation called for the development of "an integrated ecosystem conservation strategy for the Yukon and Northwest Territories, beginning immediately" (Anon., 1978).

Our research towards this objective has brought together the talents of a number of graduate students, including the author of this study, as well as the Yukon Wildlife Branch and the Canadian Arctic Resources Committee as associated research and publication arms respectively. The Donner Foundation of Canada and the World Wildlife Fund (Canada) have been principal funders, as well as the President's Committee on Northern Studies at the University of Waterloo. The project has benefitted also from the inputs of professional resource and land use planners at various institutions and members of citizen conservation organizations.

To accomplish the objective stated at the CARC National Workshop, the first research phase involved the identification of the lands worthy of preservation in the Yukon and an analysis of institutional arrangements capable of protecting each site. Thirty-five sites were identified, totalling 25 per cent of the Yukon Territory. The majority of these sites had previously been recognized by governmental parks and wildlife agencies and through surveys of the International Biological Program. One site only, Kluane National Park, had been protected in any substantial way. This phase of the research involved eleven graduate students from the departments of Planning, Geography and Biology, working with the two research directors; the results were published as a monograph entitled *Environmentally Significant Areas of the Yukon Territory* (Theberge et al., 1980). Identification and mapping was undertaken at the 1:1,000,000 scale.

The second phase of the research, part of which is reported in this monograph, was to survey ecological and human-use characteristics of chosen ESAs to carry planning to a stage where specific areas could be officially and legally recognized, i.e. boundaries established and appropriate management regimes recognized.

To accomplish this objective, a conceptual framework was developed for integrating information on biophysical and cultural values. The resource survey approach described in this report involves a three-tiered level of information analysis and comprises Jamie Bastedo's most important contribution to the ESA research (Bastedo, 1982). This approach was designed to overcome some of the perceived drawbacks of existing techniques, particularly the ecological land classification (inter-disciplinary) approach. An appraisal of the strengths and weaknesses of this and other major resource survey types is included in this study.

The method described here is a conceptual and technical contribution to the field of environmental analysis and provides a practical option for agencies with mandates to collect and analyse resource information related to environmentally significant areas. We hope that the method will transcend its Yukon-applied origins and find application to park and reserve planning elsewhere. For resource planners or managers who are unsatisfied with existing resource survey methods, we believe this approach deserves serious consideration.

The success of the Yukon Environmentally Significant Areas research program ultimately will be measured by the quality and quantity of land actually placed under preservation/protection mandates. In doing the necessary system planning and resource surveys, i.e. the science side of conserving ecosystems, we hope to act as catalysts to encourage governments to carry out their side of it - the creation of parks and reserves where they are needed. The resource survey method described in this monograph aims to accelerate this process, for in the Yukon, as elsewhere, time is our most rapidly dwindling resource.

Acknowledgements

I owe the largest measure of gratitude to John Theberge and Gordon Nelson - to John, for helping with the technical, logistical and editorial end of this research; and to Gordon, for exposing me to the realities of social and institutional concerns in the north. As well, thanks go to the ESA research team at the University of Waterloo, especially Dave Sauchyn and my wife Brenda.

While in the Yukon, my research benefitted greatly from the consultation and field support provided by the 1980 Ecological Land Survey team. Among the distinguished advisors to this team were Ed Oswald and Bill Cody with whom I struck up a fruitful correspondence. Members of the Yukon Wildlife Branch were also very helpful, especially Doug Larsen, Manfred Hoefs, Christine Boyd and Barney Smith. Too numerous to mention by name are representatives from other agencies across the country who provided me with information and advice. These agencies include the Canadian Wildlife Service, Parks Canada, the Lands Directorate of Environment Canada and the British Columbia Resource Analysis Branch.

For the production of this report, I thank Sue Sherry and Alyce Hiebert for the initial typing, Susan Friesen and Karen Steinfieldt who handled text inputting and final text preparation, Marko Dumancic for his assistance with typesetting and Gerry Boulet and Gary Brannon of the Faculty of Environmental Studies Cartographic Unit for the cartographic work. I owe Chris Bryant a special thanks for coordinating the whole production phase.

Finally, I am very grateful to the Donner Foundation and the World Wildlife Fund for their generous financial support.

Jamie D. Bastedo

TABLE OF CONTENTS

CHAPTER 3

CHAPTER 4

APPENDIX G

APPENDIX H

APPENDIX I

APPENDIX J

LIST OF FIGURES

APPENDIX G

LIST OF TABLES

APPENDIX I

APPENDIX J

Introduction

SCOPE OF RESEARCH

The intent of this research was to develop a resource survey method applicable to land use planning in the Yukon's Environmentally Significant Areas. While the outcome of research was intended for specific application to the identification and protection of national and territorial parks, national wildlife areas and ecological reserves in the Yukon, it can be applied to the planning of parks and reserves in other relatively unaltered areas. Some of the procedural details of the survey method were designed specifically for northern and alpine environments. However, the general approach is based upon concepts and principles which may be adopted for resource surveys in virtually any environment at various scales. Therefore, this research should be viewed as a contribution to the relatively new and evolving field of environmental analysis, in which the theoretical and applied bases have neither been clearly established nor broadly accepted (Cattell, 1977).

The task was to search for procedures of collecting, analyzing and presenting ecological information useful in determining policy and management direction within ESAs and in guiding delineation of their boundaries. To accomplish this task, three strategies were available: (1) to accept one or more existing resource survey methods or combinations thereof; (2) to construct a method which amalgamates the most desirable attributes of existing ones; or (3) to devise an entirely new resource survey method. From the outset, it was assumed that if a ready-made survey method was applied to ESAs, the likelihood of generating irrelevant data or needless costs would be high. Therefore, for this research, a combination of the second and third strategies was adopted.

OVERVIEW

In Chapter 1, the philosophical and technical bases for planning in ESAs are discussed in order to provide the foundation upon which the design of a suitable resource survey method can be made. In Chapter 2, the main strengths and weaknesses of existing resource survey methods are outlined in the context of six pre-defined appraisal criteria: economy, flexibility, replicability, ecological validity, communicability and applicability. Based on the results of this appraisal, Chapter 3 presents a resource survey approach which involves the independent analysis and subsequent integration of abiotic, biotic and cultural (ABC) resource information. Procedures for the biotic component (vegetation and wildlife) of the proposed approach are given in detail. Using the same appraisal criteria used in Chapter 2, an attempt is made in the final chapter to outline strengths and weaknesses of the biotic survey procedures, making critical comments on the overall ABC survey approach where relevant.

DEFINITION OF RESOURCE

A functional interpretation of the term "resource" was adopted for this research, derived largely from the work of Van der Maarel (cited by Van der Ploeg and Vlijm, 1978). Within this interpretation, natural and human resources are divided into four categories based on their functional relationship to society: (1) *production resources* (supply of matter and energy); (2) *carriage resources* (components of environment which act as carriers of human activities, artifacts and waste products); (3) *information resources* (opportunities for scientific research and monitoring, education, indicators, gene pools); and (4) *regulation resources* (species or processes controlling the integrity of ecosystems). From the beginning of this research, it was assumed that for the Yukon's ESAs, as for all areas of high environmental significance elsewhere, the latter two types of resources - information and regulation - would be of primary concern.[1] Throughout this report, references to resources and resource values in ESAs should be understood in this context.

[1] Production and carriage resources would receive greater attention during large scale, site specific land use planning studies within particular ESAs.

CHAPTER 1

RESOURCE SURVEYS AND ESA PLANNING

THE ROLE OF RESOURCE SURVEYS IN LAND USE PLANNING

To understand the special circumstances under which this research was initiated requires first, recognition of where resource surveys fit into the land use planning process in general, and second, examination of their specific role in planning for ESAs.

"Planning" is here defined as the use of scientific and technical information to provide alternatives for decision-making or, in other words, planning links knowledge to action (Steiner and Brooks, 1980). The term "land use" refers to human activities related to the utilization of resources within a spatially defined area. "Land use planning" then is the linking of information about an area of land to the action of allocating land units for various kinds and intensities of resource use. As an idealized process, land-use planning consists of ten major steps:

1. *Planning Objectives* - establish land-use planning objectives.
2. *Information Requirements* - determine broad information categories required to meet planning objectives.
3. *Data Variable Selection* - select the scope of data variables necessary to fulfill information requirements.
4. *Classification* - categorize environment into mapped land units according to a pre-determined set of criteria.
5. *Data Acquisition* - decide on the procedures, level of detail and sampling intensity for acquiring raw (unmanipulated) data.
6. *Data Interpretation* - translate raw data into a form which reflects selected resource attributes and which allows comparison of these attributes among land units.
7. *Evaluation* - based on a combination of resource attributes, determine: (a) the relative appropriateness of land units for various uses or (b) the relative worth of land units in terms of some pre-defined value measure.
8. *Prescription* - identify opportunities and constraints related to land use.

9. *Communication* - organize information into a format suitable for decision-making purposes.
10. *Zoning or Other Land Use Controls* - assign land units to policy classes pertaining to the control and allocation of resource use.

For the purposes of this research, it was assumed that a resource survey should provide the major thrust of the land-use planning process; ideally, surveyors would undertake seven of the ten steps including data variable selection, classification, data acquisition, data interpretation, evaluation, prescription and communication (steps 3 to 9). The design of resource surveys should be guided both by the earlier steps, in which planning objectives and information requirements are spelled out, and by the final step, in which resource information is applied to zoning or other land use controls. These steps are normally the responsibility of management agencies having discretionary power over the future use of land. In specifying objectives, information needs and potential applications of results, such agencies act as proponents, defining the terms of reference under which a resource survey is to be designed and carried out.

PHILOSOPHICAL BASIS FOR ESA PLANNING

As independent research carried on within the educational context of a university, this quest for a resource survey method for ESAs was undertaken in the absence of a specific proponent. Therefore, due to the wide array of possible agencies and organizations responsible for the future management of ESAs, a common ground was needed from which to formulate appropriate terms of reference for ESA planning generally. Unconstrained by particular agency objectives, a basic planning philosophy was formulated which should be acceptable to all potential managers of these lands, be they park, reserve, or wildlife agencies, native groups, or other types of organizations.

The original ESA report written by Theberge, Nelson and Fenge (1980) pointed to the necessity of adopting an ecosystem approach (after Nelson, 1980) to planning (see Appendix A). Though ESAs were originally identified on the basis of biotic (vegetation, wildlife), abiotic (geology, hydrology) and, to a lesser degree, recreational criteria, the concept of environmental significance has since been more broadly defined to include greater consideration of cultural (human artifacts and activities) resources (Hans, 1981). Therefore, given that potential managers of ESA lands recognize the interdependence of abiotic, biotic and cultural phenomena, a combined ecosystem-human ecological approach should provide the philosophical underpinning of all land use planning in ESAs.

Such a philosophy applies not only to the Yukon's ESAs, but to all land areas for which the conservation of natural values merits high priority in land use planning. In Canada, these areas could include National Parks, National Wildlife Areas, Federal Migratory Bird Sanctuaries, Ecological Reserves, Canadian Wild Rivers, Provincial/Territorial Wilderness or Natural Environment Parks (or corresponding zones) and

Provincial/Territorial Wildlife Management Areas. Common to all of these areas is a planning philosophy which incorporates the following themes:

1. **Ecological principles should take precedence over principles of economic determinism in the establishment of planning priorities.**

   Criteria for making decisions concerning resource use should emphasize ecological values above other values related to economic growth, production, feasibility and profit.

2. **Protection and preservation should be the dominant resource conservation strategies.**

   In all considerations involving the possible reduction of ecological values, the strategies of preservation and protection should take precedence over impact mitigation or post impact restoration.[2]

3. **Planning should be primarily *pro*-active rather than *re*active**

   As far as possible, planning initiatives should arise from the recognition of management opportunities rather than from the anticipation of land-use conflicts.

4. **Key areas of ecological importance should be identified and given special attention in planning.**

   As there are ESAs within a given region, so, at a larger scale, there are areas within ESAs which may require detailed site information for management purposes.

5. **Ethical considerations should play a role in decision-making.**

   Reverence for land, life and diversity (Dorney, 1978) should be the dominant ethical principle guiding land-use decisions.

----

[2] For further discussion of these different types of strategies, see Naysmith (1971).

TECHNICAL BASIS FOR ESA PLANNING

The following technical guidelines for designing a resource survey are based upon the philosophical goals stated above. In articulating these guidelines, Dorney's and Hoffman's "technical principles for environmental management" were drawn upon heavily (Dorney and Hoffman, 1979; Dorney, 1978).

Selection of Data Variables:

Data variables should reflect both structural (descriptive) and functional (relational) aspects of the environment.

Classification:

Distinct but complementary classification schemes for categorizing and mapping abiotic, biotic and cultural land units are necessary in order to elucidate the nature and degree of interaction among these three resource categories.

Data Acquisition:

Data acquisition should involve: (a) procedures which yield replicable results; (b) sufficient level of detail to allow generalization of results; and (c) a sampling intensity which captures the range of ecological diversity within an ESA.

Data Interpretation:

Raw data should be translated into a form which reflects ecological features and processes critical to ecosystem self-maintenance.

Evaluation:

Interpreted data should be combined in such a way as to enable differentiation of land units based on their relative ecological values and sensitivities.

Prescription:

Prescriptive statements should focus on management recommendations for land units of high ecological significance and on considerations related to boundaries and buffer areas.

Communication:

Results of the survey should be organized into an information package which is meaningful to park, reserve or wildlife agencies, native groups and the general public.

In Appendix B, further technical guidelines are laid out, particularly in terms of certain northern environment features and processes which should be considered in designing resource surveys specifically intended for application to arctic and sub-arctic sites.

CHAPTER 2

AN APPRAISAL OF THE MAJOR TYPES OF RESOURCE SURVEY METHODS FOR POSSIBLE APPLICATION TO ESA PLANNING

Resource surveys, in measuring the position and extent of one or more resources within discrete land units, aim to provide information pertinent to decisions on the kinds and intensities of land use appropriate for a given area. The diversity of resource survey methods currently available reflects the varied conceptualizations that surveyors have made when imposing order on nature's apparent complexity. Clearly, the possibilities for variation are infinite, suggesting that no single method is uniquely suited for application to land use decisions in ESAs. Therefore, this search for a method involved the appraisal of different types of resource surveys in order to develop procedures in accord with the philosophical and technical planning principles described in the previous chapter.

DESCRIPTIVE CATEGORIES FOR RESOURCE SURVEY METHODS

In light of the proliferation of resource survey methods throughout the 1960s and early 1970s, several researchers created frameworks to categorize and review selected methods for specific purposes. One comparative framework of note is that created by Pierce and Thie (1976) for the Canadian Government's Department of Environment. Eight examples of surveys carried out by the department were categorized into two groups: "Integrated" and "Non-integrated", the former type including "Multi-disciplinary" and "Biophysical" surveys and the latter including "Disciplinary" surveys. Although Pierce's and Thie's categories were not defined explicitly, their descriptive framework was adopted for this study because it provided the clearest and simplest schema for comparing the salient features of resource survey methods. Based on examples noted later in this chapter, Pierce's and Thie's schema was refined to include precise definitions and expanded to accommodate additional categories (Figure 1).

Resource surveys were first separated into two mutually exclusive categories based on the presence or absence of procedural integration, recognized at various steps in the survey process. For comparative purposes, three types of integration were identified:

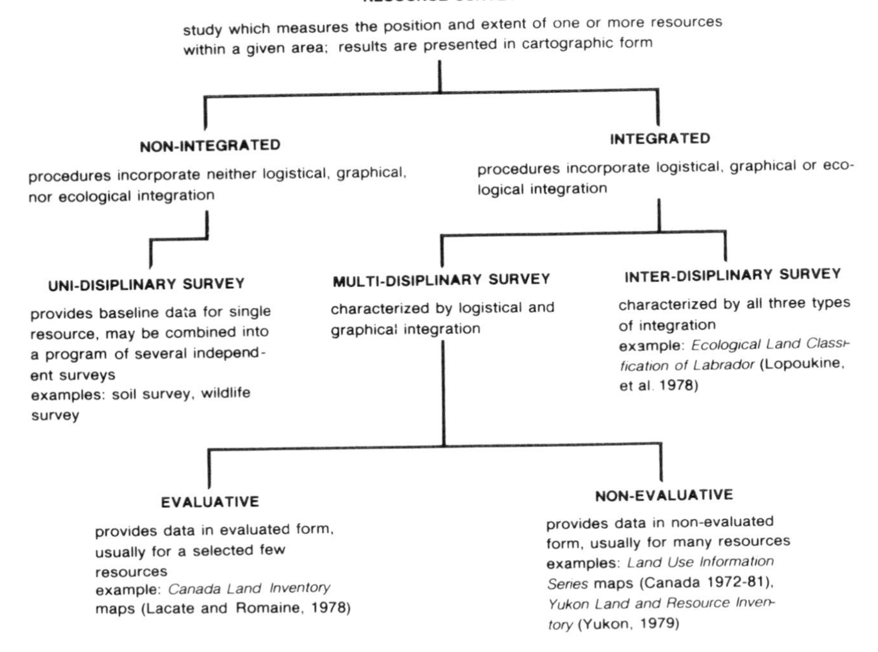

Figure 1 Descriptive Categories for Resource Survey Methods

1. *Logistical Integration* - integration of research efforts (field schedules, transportation, camps, etc.) and write-up activities among surveyors from several resource disciplines.
2. *Graphical Integration* - cartographic information from several resource disciplines is displayed separately (resource-specific legends) on: (a) a series of maps of similar scale and format or (b) on one map.
3. *Ecological Integration* - raw or interpreted data are combined to identify "ecological" units through the recognition of interrelationships among resource categories.

Further subdivision of methods was made according to the number of disciplines involved and the type of integration incorporated into the design.

Survey types categorized as *Integrated* are characterized by a research design which provides baseline data pertinent to the planning and management of several resources. Two or more specified disciplines (geology, botany, etc.) are involved using a compatible method for acquiring and presenting the data (Figure 1).

Within the "Integrated" category, *Inter-disciplinary* (Pierce's and Thie's "Biophysical") surveys represent the highest order of integration currently available in the field of environmental analysis; included are procedures which together incorporate logistical, graphical and ecological integration. Representatives from several disciplines work together as a team to produce

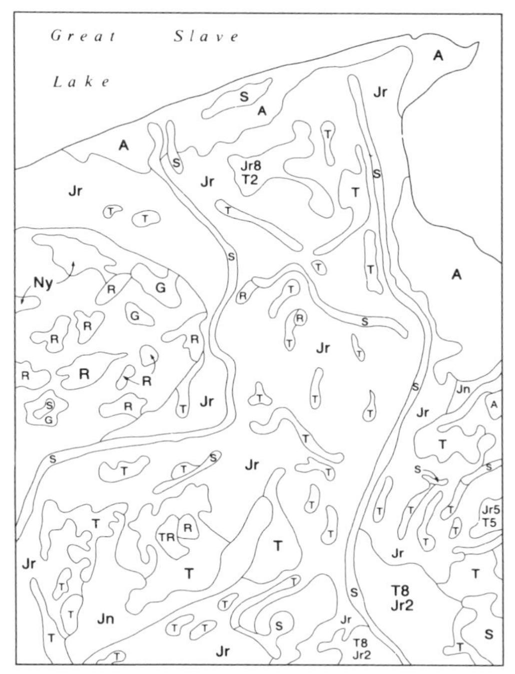

a) UNI-DISCIPLINARY SURVEY

Uni-disciplinary example shows results of a soil survey of the Slave River Lowlands, Northwest territories (Day, 1972). Colour-coded legend describes soil series/complex type, soil group, soil moisture class, dominant topography, and soil profile. For example. A=Alluvium, Cumulic Regososol, well to poorly drained, level to hummock, grayish brown calcareous loamy sand to silty clay loam. Other soils series include: T=Talcton, Jr=Jerome, Jn=Jean, R=Rock outcrop, G=Grand detour, etc. Numbers refer to relative proportion in terms of per cent (e.g. 8=80%).

1:63,360

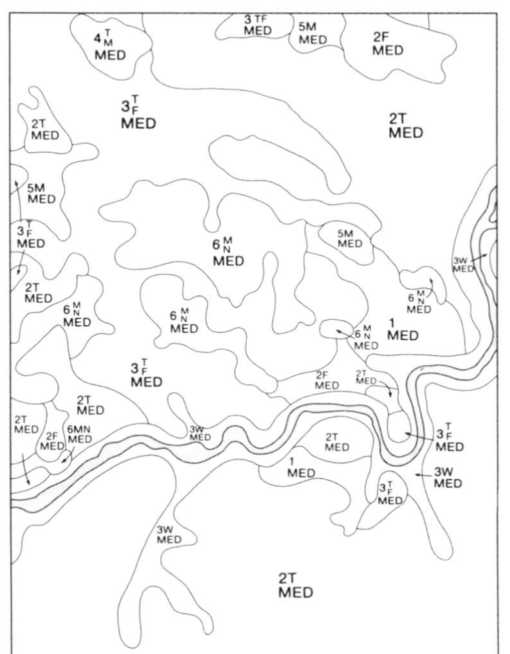

b) EVALUATIVE MULTI-DISCIPLINARY SURVEY

Evaluative Multi-disciplinary survey showing land's relative capability to support wild ungulates for the NTS map sheet 83 H, Edmonton, Alberta. The land is divided into units on the basis of physiographic characteristics judgea to be important to ungulates. Each unit is assigned a capability rating (Class 1 to 7, colour-coded) according to factors limiting habitat productivity (climate, soil moisture, fertility, topography, etc.) Examples: 1 MED = no significant limitations to production of moose, elk or deer; 6 M/N MED = severe limitations to production of same due to excessive soil moisture and adverse soil characteristics.

1:250,000

Figure 2 Comparison of Map Products for Major Resource Survey Types

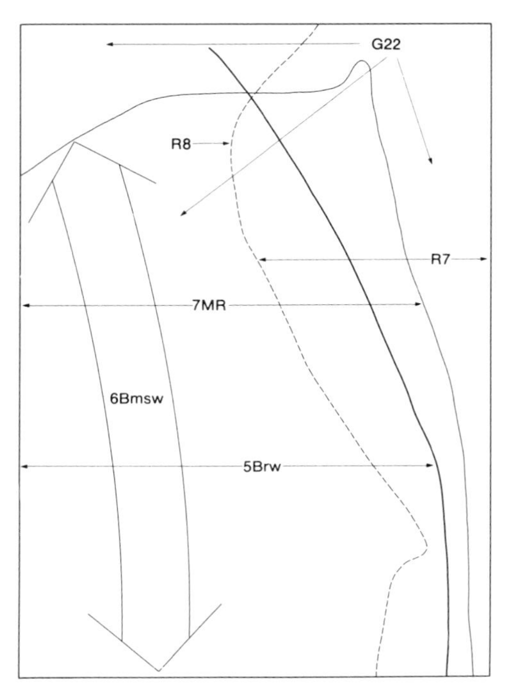

c) NON-EVALUATIVE MULTI-DISCIPLINARY SURVEY

Non-evaluative Multi-disciplinary example is taken from the Land Use Information Series (Canada, 1972-1981), NTS map sheet 106M, Richardson Mountains, Northwest Territories. Coded information on map refers to wildlife locations (5 Brw = barren ground caribou winter range; 6 Bms/w = barren ground caribou migration route, summer and winter), wildlife harvesting (7MR = caribou hunting area, October to April), and terrain evaluation for recreation (R7 and R8 = distinct recreation zones).

1:250,000

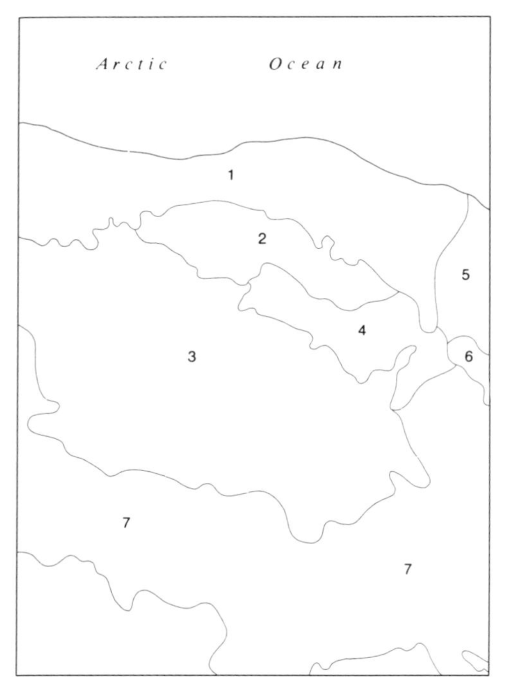

d) INTER-DISCIPLINARY SURVEY

Inter-disciplinary example is part of an "Ecodistricts" map for northern Yukon Territory (Wiken et al., 1978). The accompanying map legend describes each numbered unit with respect to recurring landforms, soils, vegetation, and water bodies. Ecodistricts are defined as "areas of land having an ecologically unified pattern consisting of various components such as: assemblages of local or, in a few cases, individual regional landforms; assemblages of soil associations; plant districts or, occasionally, plant regions, meso-large-scale micro-climate areas; and sub-basins and small watersheds."

1:500,000

a resource survey package in which *inter*-relationships between resource variables are emphasized (Figure 2d).

Multi-disciplinary surveys are distinguished from inter-disciplinary ones in that the data are not integrated ecologically. Rather, resource-specific information is presented separately on one map or a series of maps of the same scale giving a *multi*-faceted description of the resource base (Figure 1). In this way, each resource may be perceived within a unique geographic context, and be related spatially to other resources. Multi-disciplinary surveys have been sub-divided into "Evaluative" and "Non-evaluative" categories according to whether resource information is evaluated or not.

Evaluative multi-disciplinary surveys provide information in evaluated form and usually focus on a few selected resources. This approach systematically evaluates resource data in terms of land capability for various uses (e.g. agriculture, wildlife production, recreation) and then integrates them graphically onto a series of resource-specific maps (Figure 2b).

Non-evaluative multi-disciplinary surveys provide data in their raw form without an evaluation being made to compare land units for various purposes. Typically these surveys are more comprehensive than evaluative surveys, often including information related to existing and/or proposed land uses, in addition to information on natural resources. Presented in an unevaluated form, this information appears on one map, or on a series of similar maps (Figure 2c).

Uni-disciplinary (Pierce's and Thie's "Disciplinary") surveys are designed to provide specific baseline data for only one resource category (e.g. wildlife, vegetation, soils).[3] As such they are non-integrated and require expertise from a minimum of resource disciplines (Figure 2a).

CRITERIA FOR APPRAISING RESOURCE SURVEY METHODS

The prime criterion guiding the selection or design of any resource survey method is whether or not the method provides information that can be utilized to satisfy objectives pre-specified by the user, i.e. the means must be appropriate to the ends. This prerequisite must be met within the limitations imposed by such technical realities as:

1. Financial, temporal and personnel constraints.
2. The geographic and administrative scale at which survey results are to be applied.
3. The required level of precision, accuracy and objectivity in data collection and manipulation.
4. The adequacy of existing ecological knowledge.

----

[3] The term "soils" is used here in its classical sense, i.e. surface material composed mainly of decomposing organic matter and fine mineral particles.

5. Technical competence of the user(s).
6. A survey method's history of success in meeting particular objectives.

These factors were considered in determining the utility of various approaches for ESA planning needs. To this end, a set of appraisal criteria were devised to provide a pre-defined yardstick against which the strengths and weaknesses of various methods could be judged. These criteria include: Economy, Flexibility,, Replicability, Ecological Validity, Communicability and Applicability. In the following sections, these criteria are defined, their general significance to resource research is indicated and their specific requirements pertinent to ESA planning and management are noted.

Economy

Economy refers to the degree to which available funding, research time and manpower have been utilized efficiently in relation to data quality and quantity. Because of the complexity of ecological systems, the amount of data capable of being generated among resource disciplines is infinite, as are the manipulative possibilities. In designing a resource survey, a team may begin with the attitude: "Let's measure everything and decide what to do with it later" (Feachem, 1977, p.6). Such approaches are doomed to failure as they inevitably yield data which are either of marginal usefulness or are so complex as to hinder comprehension; and they are expensive. Therefore, the need for economy requires that data relevance be determined in advance in relation to the particular objectives of the users. Another important consideration related to economy is the amount of field work required. In comparison to remote sensing, mapping and data storage activities, it is field work which involves the greatest outlay of time and money.

Under these circumstances, the cost effectiveness of a survey should be weighed primarily in relation to first, the degree to which field data are conducive to extrapolation (hence reducing field time) and second, the relevance of results to the users' objectives.

Flexibility

Flexibility is determined by a resource survey's capability for application to different environments and different scales of perception related to decision-making. The former quality, *geographic flexibility*, allows for a wide geographic application of the design and for the comparison of survey results from areas which may be quite distinct ecologically. Equally important is *perceptual flexibility*, characterized by a resource survey design which is applicable to a hierarchical range of administrative levels. As the nature and scope of objectives vary for different scales of planning, so, accordingly, do mapping scales and the corresponding type and detail of data. The most flexible resource surveys are hierarchical in design to permit broad-scale analysis of a diverse study area while allowing for the identification of key sites that warrant more detailed analysis at a finer scale.

Replicability

The concept of replicability requires that repeated measurements of the same phenomenon produce consistent results despite the pressures of inherent bias and the fact that different individuals might be involved in the research (Coleman, 1977). A high level of replicability is a desirable design feature for four reasons. First, the capability for data extrapolation beyond specific sampling sites is positively related to replicability (Mitchell, 1979); second, it ensures that a strong information base is provided for future research (Cowan, 1977); third, scientifically acceptable data are an essential prerequisite to sound resource management (Cowan, 1977); and last, it increases the defensibility of survey results and recommendations, especially when public participation and/or adversary type hearings are part of the planning process (Dorney, 1978).

To satisfy this criterion, procedures for collecting and manipulating data should meet standards acceptable to the appropriate resource disciplines. Remote sensing techniques, field procedures, classification criteria and mapping rules should be explicit, documented and based on observable attributes of the environment. For any information which is inferred, e.g. successional trends or morphogenetic processes, the specific assumptions and observable evidence upon which the inferences are based should be stated. Finally, classification units should be recognizable on the ground as well as by remote sensing, and identifying characteristics should be easily measurable.

Ecological Validity

As ecological considerations gain influence in the establishment of planning priorities, it becomes increasingly important to judge the validity of their theoretical and technical foundations. Ecological validity is a function of the soundness of operational definitions, the logic behind the selection of variables and the adequacy of procedures employed to measure these variables. Broadly defined, ecological validity is the degree to which resource survey procedures are grounded on current and/or appropriate ecological knowledge.

To achieve ecological validity, survey methods must be based upon operational definitions founded on proven ecological theory and the inclusion of variables whose selection can be justified by existing ecological knowledge. As well, the method should be sensitive to human ecological processes which affect the quantity and quality of natural resources.

Communicability

Communicability refers to the ability of a resource survey method to convey information in a comprehensible format to a broad range of users of different backgrounds (Coleman, 1977). Typically these users would include politicians, resource planners and managers, ecologists and interested members of the public.

This criterion can be met if the design is constructed around concepts and logic that are explainable to both technical and non-technical audiences, maps are readily understandable and map units are compatible with the size and diagnostic features of land units familiar to resource planners and managers. As well, information from separate resource disciplines should not be "masked" or lost during classification or mapping procedures. This latter condition is particularly important for field recognition by those people interested in knowing the exact distribution of specific resources.

Applicability

Applicability is a measure of the relevance of resource survey results to land use planning. The best testimony to its relevance is the degree to which it has been used for making decisions related to land use. Does the method capture information which has aided decisions? Have the results been used and if so, how successfully? Applicability then, is a function of a method's "track record" in the arena of environmental planning.

APPRAISAL OF THE MAJOR RESOURCE SURVEY TYPES

In this section, the four resource survey types - uni-disciplinary, evaluative multi-disciplinary, non-evaluative multi-disciplinary, and inter-disciplinary - are appraised in the context of the above criteria. This appraisal is framed around representative examples of each type and around pertinent literature. Though many of the survey examples were selected from northern Canada, reflecting the original orientation of this research, they do illustrate the state-of-the-art of resource survey work in general. Sample maps and/or reports from each survey type were examined to enable comparison of the different methodologies and to permit summary comments to be made on their relative strengths and weaknesses for ESA planning. Of the four resource survey types, the inter-disciplinary approach is given the most emphasis in the discussion. It merits special attention as it is currently the most widely adopted type of survey, even though it is the most recently developed and, consequently, the least tested.

The results of this appraisal are summarized in Table 1.

STRENGTHS(+) AND WEAKNESSES(–) IN RELATION TO APPRAISAL CRITERIA		UNI-DISCIPLINARY	EVALUATIVE MULTI-DISCIPLINARY	NON-EVALUATIVE MULTI-DISCIPLINARY	INTER-DISCIPLINARY
ECONOMY	+	- isolated surveys economical for simple site planning decisions			- potential savings through logistical integration
	–	- program of independent surveys wasteful due to overlap and duplication in field work, transporation and lab work	- economic feasibility depends on pre-existing information base	- economic feasibility depends on pre-existing information base	- potentially wasteful through data "overkill" during data acquisition
FLEXIBILITY	+	- can accomodate a great variety of data		- can accomodate a great variety of data at various scales	- designed for a variety of ecosystems - hierarchical data acquisition allows broad perceptual flexibility
	–	- most surveys represent horizontal cleavages of land at fixed scales	- geographic flexibility limited by validity of capability assumptions - perceptual flexibility restricted to 1:250,000 mapping scale	- scales of perception not standardized from one resource to another	- cannot include information related to wildlife dynamics, archaeological resources, nor point and line data of any kind
REPLICABILITY	+	- standardized procedures used to collect raw data - data based on observable evidence therefore little opportunity for bias	- classification criteria based on observable evidence; assumptions for each made explicit		- standardized procedures used to collect raw data
	–			- replicability limited by geographic and temporal variations in data availability	- unspecified procedures for ecological integration of data - lack of precise standardized classification criteria - dependence on intuition
ECOLOGICAL VALIDITY	+	- suited for resource variables requiring fragmented view of environment			
	–	- generally unsuited for depicting interrelated aspects of environment	- existing spatial patterns and interrelationships between resources may be ignored - over-reliance on soils data to determine productivity - assumptions not adjusted for ecological variance between areas - human ecological data excluded	- generally unsuited for depicting interrelated aspects of environment	- strong bias towards physical features in delineating land units meant to be ecologically homogeneous; this bias obscures interrelationships between wildlife, vegetation, and abiotic components - human ecological data excluded
COMMUNICABILITY	+		- rapid comprehension possible due to low number of resources, maps with strong visual impact		- nested data facilitates computer storage
	–	- data often presented in technical form - program of independent surveys results in inconsistencies in report styles, levels of detail, and mapping scales and legends	- potentially useful raw data are lost	- potentially confusing due to disjunction or overlap of several resource variables	- large number of characteristics per map unit - inconsistencies in terminology - user training required - potentially useful raw data are lost
APPLICABILITY	+	- applied to simple planning problems on an ad hoc basis		- used as initial familiarization and reference tool for: production and/or review of environmental impact assessments of major projects, road/pipeline route evaluation and planning, administration of territorial land use regulations, and education	
	–	- difficult to apply to broad land use planning decisions	- though applications many and diverse, most related to resource production or environmental impact determination rather than resource protection and preservation	- problems of inappropriate scale, level of detail, and need for updating occasionally limit potential applications	- majority of data has found minimal application in broad land use planning decisions - failure to derive practical meaning from single map product an important factor in restricting applications

Table 1 Summary of Appraisal of Major Resource Survey Types

UNI-DISCIPLINARY SURVEYS

The resource surveys of this type are so diverse as to make generalizations about the category as a whole difficult. Therefore, among the studies examined, four were selected which illustrate collectively the general strengths and weaknesses of the uni-disciplinary approach. The studies reviewed include two which represent isolated uni-disciplinary surveys (*Reconnaissance Soil Survey of the Takhini and Dezadeash Valleys in the Yukon Territory* (Day 1962), and *Population Ecology Studies of the Polar Bear in Northern Labrador* (Canadian Wildlife Service, 1980)) and two which represent a program of independent uni-disciplinary surveys, the environmental and social research program associated with the MacKenzie Valley and Northern Yukon pipeline proposals (Department of Indian Affairs and Northern Development, 1972; Crampton, 1973; Lavkulich, 1973) and the renewable resource research program in Kluane National Park (Theberge, 1972, 1974; Krebs and Wingate, 1976; Hoefs, 1976; Douglas 1974).

Strengths of Uni-disciplinary Surveys for ESA Planning

The obvious simplicity of the uni-disciplinary approach makes it attractive from the standpoint of economy when nothing more is needed than the survey of one specific resource. A minimum of personnel is required and the fieldwork usually can be accomplished in concentrated periods of research undertaken by small teams, commonly of only two people, as demonstrated by Day's soil survey and by most of the polar bear work in Labrador. Simple planning decisions for large scale, site-specific projects can be met economically using information from such isolated surveys.

The diversity of subject matter in uni-disciplinary surveys which have been undertaken in the north attests to the geographic flexibility of this approach. Given the necessary expertise and funding, there are few resource variables of the northern environment for which a uni-disciplinary survey could not be designed. For example, specific uni-disciplinary surveys have been completed on permafrost-terrain relationships, earthquake hazard zones, eagle nests, seasonal distribution and migration patterns of caribou herds and archaeological and historic sites.

Standardized and validated procedures are the rule for uni-disciplinary surveys since, through a long history of application, much progress has been made in refining univariate sampling methods (Frayer et al., 1978). There were no exceptions among the examples reviewed. For example, the polar bear survey was conducted by qualified wildlife biologists using well established techniques for studying the movements and distribution of large mammals. In this case, the conferred replicability provided a strong information base from which to conduct future research and monitoring studies.

In every case, resource data were based upon directly observable evidence from the field and/or from aerial photos, and they were presented in a raw, uninterpreted form. In this way, the introduction of subjective measures was avoided, thereby minimizing the possiblity of research bias and maximizing the possibility of replication.

With respect to ecological validity, most uni-disciplinary surveys foster a fragmented view of the environment in that it is described on a resource-specific basis. Though unsuitable for analyzing interrelated aspects of the environment, this approach is suitable for meeting other information requirements which necessitate the analysis of resources within a fixed spatial and/or temporal context. Examples from among the studies reviewed included the identification of summer concentration areas for polar bears and the identification of archaeological and historic sites.

In addressing relatively simple planning problems related to parks and reserves, the uni-disciplinary approach has been applied successfully. For instance, in National Parks, isolated uni-disciplinary surveys have provided information on an ad hoc basis to resolve specific conservation issues (Cattell, 1977).

Weaknesses of Uni-disciplinary Surveys for ESA Planning

Programs of independent uni-disciplinary surveys involving several teams of resource specialists have been carried out for broad land-use planning of large areas. This approach was used for management planning at Kluane National Park and for routing and mitigation considerations associated with the MacKenzie Valley/Northern Yukon research. Both programs involved the establishment of numerous field camps separated from each other spatially and temporally, the separate scheduling of aerial surveys and access support and independent efforts during remote sensing and mapping procedures. It was only due to a generous government budget that these programs could be carried out in this fashion. Without considerable and guaranteed financial backing, such an endeavour would not be feasible; even with this backing, the endeavour may be economically wasteful due to the potential for overlap and duplication in fieldwork, transportation, support staff and lab and cartographic work.

A major weakness of such programs is that attempts to make broad land use decisions from the results of several surveys are hampered by inconsistencies in the presentation of data. Common to both the Kluane and MacKenzie Valley/Northern Yukon research are a wide array of report styles, levels of detail and mapping scales and legends. Size and location of the study areas also varied among surveys resulting in uneven overall coverage. Under these circumstances, potential conflicts, constraints and opportunities offered by the resource base as a whole are difficult to discern.[4]

----

[4] Moves to integrate such efforts involving multiple resources no doubt

Though most uni-disciplinary studies are carried out to provide information which will eventually be used in land use decisions, they commonly serve the parallel function of scientific research, especially in northern environments where there is much to learn. As a result, survey reports and maps are often presented in a format understandable only to other resource specialists. In reference to soil surveys in the Yukon, Tarnocai (1979) wrote that some reports "seem to be written for other pedologists and are not slanted to specific users" (1979, p.117). In several of the MacKenzie Valley/Northern Yukon survey reports, information assembled for scientific purposes was not distinguished from that which was more relevant to planning decisions. For the Kluane studies, interpretations of technical data were kept to a minimum; in the instance of an ornithological survey (Theberge, 1974), management implications of the results, by contract requirement, were removed completely from the survey report before publication.

Attempts to integrate, either ecologically or graphically, the results of several uni-disciplinary surveys have limited success due to inconsistencies in field procedures and mapping format. When surveys of this type are combined, each resource may be focussed upon without concern for interrelationships or the state of interdependency existing among other environmental components (Department of Indian and Northern Affairs, 1977). For the Kluane and MacKenzie Valley/Northern Yukon research programs, a wealth of data was generated by individual surveys. However, efforts to interrelate these data for research or planning purposes were negligible.

Another common weakness of uni-disciplinary surveys is that they usually do not incorporate perceptual flexibility into their design. Though soils and vegetation can be surveyed according to hierarchical categories corresponding to various scales of perception,[5] uni-disciplinary surveys, on the whole, represent "horizontal cleavages" of the land under investigation (Department of Indian and Northern Affairs, 1977).

Given the recent trend toward multiple resource planning, the uni-disciplinary approach has found decreasing favour among land use planners. For complex planning problems involving more than one resource, a series of independent uni-disciplinary surveys is not the solution. This approach has proven especially difficult to apply to the variety of decisions involving spatial allocation of land uses. The applicability of results from the MacKenzie Valley/Northern Yukon survey program is not in question; this program was geared more toward environmental impact assessments than

----

provided impetus for the evolution of multi- and inter-disciplinary survey methods.

5 The Canadian taxonomic classification of soils includes six hierarchical categories: order, great group, subgroup, family, series and type. Traditional categories for vegetation include: community, association and stand.

land use planning.[6] Evidence from the master planning process at Kluane however, suggests that resource information obtained from uni-disciplinary surveys had less than optimal influence on decision making.

EVALUATIVE MULTI-DISCIPLINARY SURVEYS

Most representative of this approach are the "Land Inventory" programs undertaken during the 1960s and early 1970s by the federal and Ontario governments. Since the two programs, Ontario Land Inventory (OLI) and Canada Land Inventory (CLI), were conducted co-operatively, their terms of reference and procedures were basically the same. Thus, only the more widely known CLI program is referred to in the following discussion.

Strengths of Evaluative Multi-disciplinary Surveys for ESA Planning

A major strength of evaluative multi-disciplinary surveys is communicability. Typically, these surveys culminate in a series of maps which are relatively easy to understand. As well, the low number of resources involved allows the user to comprehend rapidly the evaluated data without fear of information overload. Consistent mapping scales and explicit legends contribute to the communicability of the maps. CLI maps include a colour coding scheme for capability classes in addition to an alpha-numeric legend so that visual impact is maximized.

The replicability of the evaluative multi-disciplinary approach merits attention because of its emphasis on the evaluation stage of the resource survey process, the stage at which bias can most likely enter into the results. Most often this is due to the "apples and oranges" quandary faced by researchers who must weigh qualitatively the values of unrelated resource variables in order to assign a capability rating to an area of land.

Much literature has been published on the CLI methodology in an attempt to standardize the classification criteria and thus minimize the introduction of bias. These criteria were created by resource specialists and for each, recognition features are derived from directly observable evidence from air photos or the ground and/or assumptions are explicitly stated.

----

[6] For a summary of how this information was used, see Zoltai (1979).

Weaknesses of Evaluative Multi-disciplinary Surveys for ESA Planning

The picture created by evaluative multi-disciplinary surveys bears little resemblance to existing environmental conditions. Instead, the complexity of present ecosystems is translated into parametric land units whose boundaries are determined by the land's estimated potential to sustain future use for selected resources. In the CLI program, ecological considerations were used to delineate, on separate maps, areas significant in terms of their production capability for agricultural crops, timber and wildlife (ungulates and waterfowl). As well, areas of land significant for recreation capability were differentiated on the basis of environmental information which was relevant to recreational opportunities and present recreational use.

Four aspects of the CLI program have negative implications for the ecological validity of evaluative multi-disciplinary surveys.

First, the parametric approach to land classification creates a simplistic picture of the environment which is difficult to view in the context of ecology. By following this approach, existing spatial patterns and interrelationships between resource variables may be ignored during the classification process.

Second, in relying almost exclusively on soils information, CLI researchers may have overlooked other more important factors which determine the productivity of renewable resources. Rogers (1974) cites this as one of the major shortcomings of the CLI program, particularly with regard to wildlife.

Third, related to the latter point is the fact that many of the assumptions made concerning the productivity of renewable resources may not be valid when applied over a wide geographic area. For example, habitat features, or combinations thereof, were evaluated according to assumed preferences of wildlife. However, from one area to another, habitat preferences could differ significantly in response to variations in climate or to varied pressures from competition, predation or hunting. Similar comments could be made concerning the variations in the ecology of forests and agricultural crops.

Last, except for present recreational land use, no human ecological components or processes are incorporated into the classification scheme. All of the classifications were based strictly on the physical capability of the land. Social and economic factors such as distance to markets, location, cultural patterns, size of units and present state of the land were not considered (Lacate and Romaine, 1978).

For those users interested in the ecological information upon which the CLI evaluations were based, the resultant maps, however colourful or explicit, are of little value. By presenting only the evaluated information, the original raw or interpreted data were lost. For land use planning, these data are potentially useful particularly to resource specialists who need to recognize classified units in the field.

Though the CLI mapping program is untested in northern

environments,[7] its broad coverage across other parts of Canada suggests that evaluative multi-disciplinary surveys potentially could accommodate data from a variety of ecosystems. According to a CLI handbook, the approach adopted for the program was designed not only to be suitable for a broad geographic area but also:

> to provide an information base for resource and land use planning by all levels of government - federal, provincial, regional or local. (Environment Canada, 1978, p.6)

However, in regard to this purported perceptual flexibility of design, Mitchell cautions that

> it must always be recognized that the capability estimates are based upon a series of assumptions which may be valid at a national level but less valid in specific regions. (Mitchell, 1979, p.67)

As well, perceptual flexibility is restricted to a narrow range of administrative levels which can use data fixed at a scale of 1:250,000.

The area covered by evaluative multi-disciplinary resource surveys is typically very large, with data being accumulated and mapped by several resource agencies. The boundaries of the CLI survey embrace all of Canada's major settlement areas plus adjacent areas, totalling approximately 1,000,000 square miles. To cover such a large area, heavy reliance was put upon remote sensing with relatively few field checks and with much extrapolation based on existing data. Since its inception in 1962, the CLI program has cost many millions of dollars for photo interpretation, compilation of existing information, mapping and data storage, report publications and administrative salaries. The program was made affordable only through the diffusion of costs through various government agencies across the country.

If this approach were applied to smaller areas on the scale of ESAs, the cost would be proportionately reduced to a seemingly attractive level. The choice of this approach however presupposes the existence of enough baseline data to make inferences on resource capabilities. The CLI program was made possible through previous work in the classification and mapping of soils, collection of climatic data, studies of land use and compilation of statistics on productive capacity (Environment Canada, 1978). Not nearly enough information is available for many parts of northern Canada to make such a project economically feasible. To cover an ESA completely using the evaluative multi-disciplinary approach, the cost of acquiring the requisite baseline data would be prohibitive.

----

7 Uni-disciplinary capability studies for agriculture have been conducted in northern Canada. For example, see Rostad et al. (1979).

Results from the Canada Land Inventory have been successfully applied to a variety of planning situations, particularly in delimiting important agricultural lands, in identifying recreational opportunities and in contributing resource information to environmental impact studies.[8] In general, the CLI program has proven itself a useful tool for the purpose for which it was designed, namely to facilitate long range land use planning with respect to potential resource production and intensive management. However, these concepts upon which the program's design was based are foreign to the primary goals of ESA planning: protection and preservation of existing resources. Though the diversity of applications of the CLI information suggests that the evaluative multi-disciplinary approach provides a panacea for land use planners, a more conservation-oriented, cost effective and ecologically valid approach is needed for ESAs.

NON-EVALUATIVE MULTI-DISCIPLINARY SURVEYS

Three examples from northern Canada were chosen for the appraisal of this approach: the *Land Use Information Series* (LUIS) project (Canada, 1972-1981), *Yukon Land and Resource Inventory* (Yukon, 1979), and a pilot resource survey undertaken by the author at Frances Lake, Yukon (Bastedo, 1979).

Strengths of Non-evaluative Multi-disciplinary Surveys for ESA Planning

One of the most attractive features of non-evaluative multi-disciplinary surveys is that they can accommodate virtually any type of resource data which are spatially defined on a mappable scale. LUIS maps, for instance, portray a diversity of information at different scales, from fish spawning grounds of a few hundred square metres to registered trapping areas several square kilometres in size. The consistent success at mapping almost the entire area of the Yukon and Northwest Territories attests to the adaptability of the approach to various northern environments.

This flexibility of the non-evaluative approach is reflected in the diversity of uses to which such surveys have been applied. Kerr summarized applications of the Yukon Land and Resource Inventory and the LUIS maps:

> The Inventory is used by consultants and government mainly as a reference source to aid in bibliographic searches. The mapped information allows the researcher to assemble quickly what is known about a geographic area, and has proven to be

----

8 For a more complete summary of CLI applications, see Mitchell (1979) and Lacate and Romaine (1978).

> a useful and valuable tool in this regard . . . With respect to the Northern Land Use Information Series, our office has found it useful in providing a broad overview of land use and environmental information. In . . . reviewing environmental impact assessments of major proposed developments in the Yukon . . . the maps have been used as preliminary background information for identifying areas of potential environmental conflict. (Kerr, 1981)

These surveys have also been applied successfully to: administration of territorial land use regulations, regional planning exercises, northern education at high school and university levels and familiarization of local resource management officers (Kananaskis Centre, 1979).

Weaknesses of Non-evaluative Multi-disciplinary Surveys for ESA Planning

With respect to the criterion of economy, projects within this resource survey type exhibit a spectrum of dependency on field work corresponding to the availability of data. Among the examples chosen, the Yukon Resource Inventory was strictly a desk exercise, constructed as a bibliographic reference for existing resource data and required no original field work. In contrast, the survey at Frances Lake required a season's field work in addition to interviews and literature research. Between these extremes lies the LUIS mapping program.

For covering an equivalent area of land, the cost of all of these projects was related, for the most part, to the proportion of required field work to desk work. If the baseline data were available to the level of detail required for ESA planning, non-evaluative multi-disciplinary studies could be executed with a relatively low expenditure in time and money. However, if these data were not available, effective execution would require much costly field work, rendering the approach uneconomical for ESAs in remote areas where information is limited.

Where relevant data do exist, their type, quality and quantity will inevitably vary from one area to another, or, for the same area, from one time to another. This is especially true in the north where there has been a great disparity in research effort for different areas, and where some land dispositions are characteristically ephemeral (e.g. mining claims, land use proposals). The availability and reliablity of local information also varies. During the Frances Lake survey, the author was fortunate in contacting several local residents who could contribute valuable resource information. Presumably, these favourable research conditions would be rare for many parts of the Yukon so that the consistency of survey results for different areas would be affected. All of these factors imply that a high degree of replicability for non-evaluative multi-disciplinary surveys could be assured only across areas in close proximity or, for one area, over short periods of time.

With respect to flexibility, it should be noted that the non-evaluative multi-disciplinary approach is generally not hierarchical in the sense that separate resource features are surveyed and mapped at pre-determined scales of perception. Rather, the detail of the available information and/or the nature of the features themselves dictate the scale at which they are perceived. This design feature implies that, by using this approach, data may be difficult to standardize according to a particular scale (or scales) of perception. Thus, systematic data interpretation and evaluation would be impeded, diminishing the utility of survey results.

The possibility of confusion detracts from the communicability of maps from non-evaluative multi-disciplinary surveys. Too many resource variables may be presented to permit ready comprehension of the data as a whole. For the LUIS program, information was compiled onto one map and, for the Yukon Resource Inventory, onto a series of maps. In the former example, there is much graphical overlap among variables, thus impeding their differentiation. In the latter example, variables are disjointed graphically making it difficult to perceive their spatial relationships. A flexible overlay system of maps was considered for the Frances Lake results to provide a compromise solution to these problems.

Confusion might also arise from the inability of users to extract meaning from data which are presented in a non-interpreted and non-evaluated form. In spite of the fact that terms associated with basic land use and renewable resource data are familiar to most users, the implications for resource use may remain obscure.

Comments from users of the LUIS map series suggest that potential applications of non-evaluative multi-disciplinary surveys may be limited by problems of inappropriate scale or level of information detail. Zoltai (1981) reported that land use managers consider its scale to be far too small to show information relevant for land use management.[9] However, as with many other small scale maps, its main use is for broad, "regional" planning, rather than site specific management. In addition, data variability through time, especially among human-related resource categories, has created a demand for updating of mapped information.

A clear weakness of non-evaluative multi-disciplinary surveys relates to their ecological validity. In general, the variables included in these surveys show the distribution of existing resources and related land uses. The resultant picture of the environment is one in which elements of the landscape are juxtaposed together in an unrelated fashion. As such, surveys of this type could not fulfill ESA planning objectives relevant to functional aspects of the environment. Instead, this approach is better suited to fulfill objectives necessitating a component-oriented portrayal of the environment.

----

9 Another source (Kananaskis Centre, 1979) suggests that users are generally content with the 1:250,000 scale. Moreover, even smaller scale maps (1:500,000) have been proposed for certain portions of the high Arctic. Therefore, applicability in this case is a question of matching mapping scale to planning objectives.

INTER-DISCIPLINARY SURVEYS

Representing a synthesis of uni- and multi-disciplinary methods, the inter-disciplinary approach has its origins in the classification schemes developed by Christian of Australia (1959) and Hills of Ontario (1961). These early undertakings attempted to provide a framework for the integrated survey of resources in usually large, not easily accessible and little known areas. Lacate (1969) advanced this work by developing a system of "Biophysical Land Classification". The aim of this system was to describe and classify "ecologically-significant" units of the landscape (Lacate, 1969). Each unit, recognized at distinct topographic scales, was assumed to have a certain internal homogeneity and functional integrity (Rowe, 1979). Once delineated, these "iso-ecological units" (Jurdant et al., 1974) were to comprise the ecological basis for land use planning involving future management of lands for forestry, agriculture, recreation, wildlife and water yields.

In Canada, the inter-disciplinary approach to resource survey has undergone many modifications and refinements since 1969. Under the co-ordination of the Canadian Committee on Ecological Land Classification (CCELC) the approach, as it has developed, has been referred to variously as Biophysical Inventory (Holroyd et al., 1979), Biophysical Land Survey (Jurdant et al., 1974), Ecological Land Classification (Zoltai, 1979) and Ecological Land Survey (Environment Canada, 1980). Within the literature, nomenclature for hierarchical categories of map units has varied considerably.[10] As well, the scope of inter-disciplinary surveys has been expanded to stress interactive rather than just descriptive characteristics of the environment. In spite of these changes, the major conceptual and methodological aspects of the approach have remained basically the same.

Material for the following appraisal of the inter-disciplinary approach was drawn from the extensive literature on the subject, written by both CCELC members and their critics. Comments are made concerning Parks Canada's experience with inter-disciplinary surveys, with special reference to the survey at Banff/Jasper National Park. Also included are observations from the author's experience working with an Ecological Land Survey (ELS) team in the Yukon during the summer of 1980 (Yukon, 1980).

Strengths of Inter-disciplinary Surveys for ESA Planning

Proponents of inter-disciplinary surveys claim that, in comparison to other resource survey types relying heavily on field work, inter-disciplinary surveys result in proportionately less costs to collect the equivalent amount of data, due to savings accruing from logistical integration. Theoretically, for a relatively inaccessible area of up to 3,000 square km, a team of resource specialists and their technicians can complete the field work for

----

[10] For example, contrast Lacate (1969) with Environment Canada (1980).

intermediate scale surveys (1:100,000 - 1:250,000) within one season. Preparation for field work and post-field analysis and write-up bring the total work time to about one year. These estimates are based on the projected schedule of the 1980 Yukon ELS team.[11] This time frame will naturally vary in relation to the area covered and the level of detail specified in the terms of reference. In National Parks, for example, six years is the average time required for larger scale (1:10,000 - 1:50,000) inter-disciplinary surveys (Cowan, 1977, Appendix III).

Though the principles and guidelines of the inter-disciplinary approach are more or less standardized, a wide range of choice among operational definitions and procedures allows for the accommodation of data from virtually any natural environment.[12] In national parks for example:

> Some latitude is . . . accorded in the interpretation of both the land classifications and the specifications within the inventory variables to coincide with the diversity of natural environments occurring in the national parks system. (Cattell, 1977, p. 72)

Outside of national parks, the inter-disciplinary approach has been most widely adopted for collection of baseline data in northern Canada. This reflects the CCELC's original mandate to design a survey method intended for areas not covered during the CLI program (Gimbarzevsky, 1978). Major projects include *Ecological Land Classification of Labrador: A Reconnaissance* (Lopoukhine et al., 1977), *Ecoregions of Yukon Territory* (Oswald and Senyk, 1977), *Ecological Land Survey of the Northern Yukon* (Wiken et al., 1978) and the Yukon's Ecological Land Survey Program (Yukon, 1980).

One of the often promoted features of the inter-disciplinary approach is its capability for surveying resources at a wide variety of map scales.

> The survey is often two-pronged: synoptic information for large areas, to assess implications over a large area; and detailed information for specific sites, to answer site-related problems or to study representative ecosystems or relationships identified by the broad survey. (Environment Canada, 1980, p.7)

This perceptual flexibility is a design feature which exactly suits the terms of reference for ESA planning.

Though Rowe (1979) has described the inter-disciplinary approach as more an art than a science, this viewpoint does not apply to the remote sensing and field procedures employed during the data acquisition stage. In

----

11 As the finished report for this project was still pending almost two years after the project began, I believe that this time frame may represent an underestimation of required time.

12 For application of the inter-disciplinary approach to the urban environment, see Wiken and Ironside (1977).

most cases, raw data are acquired for each resource, using scientifically acceptable techniques, reflecting the approach's origins from standardized uni-disciplinary surveys.

An often promoted feature of the inter-disciplinary approach is the production of a single cartographic product, offered as an alternative to a potentially confusing array of single-theme resource maps. Data from several resources are presented in "nested form" (Day, 1978) within each map unit, facilitating computer storage and providing, supposedly, almost unlimited interpretive or evaluative potential. The data can be manipulated according to several points of view, e.g. "land capability (forestry, fauna, recreation), constraints, tolerances, susceptibilities, hazards, potential activities, and variable dynamics" (Department of Indian and Northern Affairs, 1977, p.13).

Weaknesses of Inter-disciplinary Surveys for ESA Planning

The greatest weakness of the inter-disciplinary approach lies in the lack of systematic, and thus, replicable methods for classification in general and for ecological integration in particular. Throughout the literature, there is universal agreement that the ordering of the environment into classification units depends upon the recognition of "ecologically significant segments of land surface". Further, these units are supposedly recognized by integrating ecological data from the various land components. Beyond this, however, there exists a serious deficiency in operational guidelines. Bossort (1978) expresses this deficiency well:

> It should be stressed that what constitutes an ecologically significant segment of land depends on a very private personal viewpoint, i.e. intuition. There is no formal explanation. As well there is no consistent, recognized theory/model, i.e. conceptual framework, which constitutes the *rules* of integration for the classification system. (Bossort, 1978, p.1)

The following excerpts typify ambiguities found in the literature:

> To what degree do areas of land have to be similar before they constitute a discrete land ecosystem? In part, the answer, like beauty, rests in the eye of the beholder. (Environment Canada, 1980, p.13)

> . . . this particular kind of . . . classification must draw its inspiration and guidelines from the observed and inferred interconnectedness of forests with soils, land forms, climate. (Rowe, 1980, p.19)

> Procedures [are dependent] on experience, intuition and empathy with the subject matter, learned predominately by

> association with an acknowledged master through osmosis . . . (Rowe, 1979, p.25)

> Many people look upon land classification as a magical process . . . Slowly we are becoming aware that this is not true, that land classification can take any form we specify. Classification systems are simply contrivances of people, structured to suit their needs, reflecting the development of the particular science at that point in time. (Walmsley and Van Barnveld, 1979, p.9)

As implied in the latter quote, the art/science of inter-disciplinary methodology has been in a state of flux due primarily to its recent origins. In the quest for standardization from project to project, the underlying concepts, logistical arrangements, interpretation and evaluation strategies, data storage systems and mapping formats have become increasingly similar. However, procedures for the most crucial step - classification - upon which the replicability of the inter-disciplinary approach must be judged, have nowhere been clearly specified nor been scientifically validated.

The need for replicability in classification has been recognized by Rowe:

> At each level or scale of mapping, practitioners need to spell out the criteria used in delineating the land units. Unless this is done, the potential user cannot judge the utility of the maps for his purposes. Nor can other workers in the field learn and build on the methodology. (Rowe, 1980, p.20)

Most important would be the criteria relating to ecological integration - the integration of biological and physical data - but in none of the studies reviewed were these criteria "spelled out". Rather, one finds only vague directives. For example:

> . . . the biological phenomenon must be checked against physiographic features to assure that boundaries have real significance. (Rowe, 1980, p.20)

In the Banff/Jasper survey, map units derived from physical features were adjusted to "more accurately reflect vegetation types" (Holroyd, 1980b, p.43), but this was accomplished in an ad hoc site-by-site fashion for which technical procedures were unspecified.

An important factor determining the relative cost of inter-disciplinary surveys is the degree to which data can be generalized during the classification and mapping process. Most inter-disciplinary surveys adopt a remote sensing/field checking strategy for data acquisition. Field time is minimized by extrapolating from sampled sites to other units which share similar recognition features (predictive approach). However, some surveys depend on the more expensive sampling strategy in which data are collected for each land unit (incorporative approach), particularly those surveys including wildlife. This is the case in Banff and Jasper National Parks where a wildlife survey has incorporated quantitative data on the seasonal abundance, distribution and habitat characteristics of all species of mammals, birds, amphibians and reptiles into all of the pre-classified biophysical map units. The need for such a large scale project has been questioned by Cowan (1977):

> Nowhere have I seen a statement of what data are required for what purpose in either Banff or Jasper. I doubt that some of the information specified can be obtained. I doubt also that some of it is needed for all species, over the total area of the two parks at this time. I am concerned that the timing of the biophysical inventory in the interest of maintaining a smoothly functioning system, not deflect the wildlife efforts from where they should be. (Cowan, 1977, p.28)

Implicit in his remarks is the fact that a less detailed, more predictive approach in surveying might be adequate, allowing more time and money for survey work in key areas in need of special attention.

Though inherently suited for remote environments, the flexibility of the inter-disciplinary approach should be qualified in that the types of data that can be accommodated are restricted to those which can be depicted in one set of geographically-fixed classification units. In this regard, the inclusion of wildlife information has proven to be an especially difficult and recurring problem. Describing the Parks Canada experience, East et al. write that:

> Difficulties arise in that different faunal groups have varying mobility, perceive the environment at multiple levels of resolution, and have behavioural patterns that alter seasonally. Researchers also have differing ideas of what constitutes a habitat. (East et al., 1979, p.215)

Other types of information that are not readily depicted in these units include archaeological resources, point and line data of all types and socio-economic characteristics (East et al., 1977).

These shortcomings of inter-disciplinary surveys in accommodating particular kinds of data have negative implications for the approach's ecological validity. In theory, land units delineated using the inter-disciplinary approach are meant to represent discrete ecosystems. Wiken and Welch suggest that:

> In characterizing any land ecosystem, one attempts to *TRAP* the essence of each by describing the:
> T - things or components present
> R - relationships of components
> A - abundance of components
> P - pattern of components. (Wiken and Welch, 1979, p.44)

In practice, however, these land units may, in many cases, have little valid relation to ecological realities. Coulombe provides a possible explanation:

> I think we have learned that the conceptual integration of structure and function within an ecological system is possible . . . I think we have also discovered that applying that conceptual understanding of the interactions of components and processes to a specified region of space and time in the real world is a monumental task. (Coulombe, 1978, p.8)

In addressing this task, the inter-disciplinary approach has had dubious success. This weakness is discussed below with particular reference to Wiken's and Welch's concepts of "Things" and "Relationships".

With respect to "Things", most inter-disciplinary surveys fall far short of capturing the "essence" of ecosystems because of the exclusion of data on existing wildlife. With regard to this deficiency, such surveys are better thought of as "*Botano*-physicals" rather than "*Bio*-physicals". Rather than including faunal information among the criteria for distinguishing one land unit from another, the trend has been to disregard this information and evaluate each land unit in terms of potential productivity, i.e. to estimate the capability of land units to support or produce wildlife (Department of Indian and Northern Affairs, 1977) or to collect only superficial data such as incidental animal observations or habitat assessments and present this information on a separate overlay. The more detailed incorporative technique being used for the Banff/Jasper survey may become popular among inter-disciplinary enthusiasts. Still, the technique does not include wildlife in the characterization of ecosystems. At Banff/Jasper, wildlife had a negligible role on setting land unit boundaries.

Inter-disciplinary surveys usually overlook another defining characteristic of ecosystems: human land use and related activities. To be truly ecological, as is claimed, the approach would describe these factors and their influence on the natural environment. Though some authors have called for more attention to human ecology (Thie, 1974a; Cowan, 1977), they represent a small minority; from its inception, the inter-disciplinary approach has been perceived as a tool for physical and biological resources only. The understanding is that inter-disciplinary surveys are meant to be supplemented by human ecological data derived from separate surveys.

"Relationships" is an inclusive term covering ecological data associated with systems (e.g. hydrological regimes, energy flows), time (e.g. plant succession or hydrological changes throughout the year) and processes (e.g. cryoturbation or mass wasting). In short, the term refers to functional relationships as opposed to structural components.

Rowe emphasized the importance of this concept in determining boundaries:

> The appeal to relationships rather than to climate or to geomorphology or vegetation alone provides the common basis for land classification... Having established such relationships,

> map lines can be drawn and extrapolated at the appropriate scale by using the observable controlling physical features. (Rowe, 1980, p.20)

Though endorsing "a shared focus on those relationships", Rowe revealed in the same article what appears to be an inordinate dependence of the inter-disciplinary approach on physical features (primarily physiography and surficial geology) in identifying ecosystems. Advocates commonly justify this bias in terms of the greater stability conferred to the data base, relative to vegetation or wildlife. It may be instead that this practice originated from the convenience of relying on physical features for remote sensing or mapping purposes.

In any case, by making physical features the "cornerstone of map unit definition" (Department of Indian and Northern Affairs, 1977), an ecologically unsound foundation is laid from which relationships are difficult to discern. Map units which are meant to reflect ecological functions instead become artifacts of a compromise between convenience and ecological validity; at best, the physical features characterizing these units *typify*, rather than *identify*, ecosystems (Alvis, 1978).

Designed with the user in mind, the cartographic products of inter-disciplinary surveys have not been as readily received as might be expected. For one reason, the complexity of resource data contained in each map unit (ten to twenty characteristics per unit) has resulted in equally complex presentation formats, sometimes difficult to understand. In the early years following Parks Canada's adoption of the inter-disciplinary approach, information was often poorly organized and presented and sometimes illogically classified (East et al., 1979). Line maps with bewildering legends were common. More recently, progress has been made in developing colour coded maps with more effective legends, particularly for the surveys of Auyittug and Banff/Jasper National Parks. Problems have also arisen from the use of terminology often unfamiliar to the majority of Parks Canada staff. As a solution, Cowan (1977) proposed that a series of special workshops be held to ensure that all those who must understand the system in order to use it effectively become thoroughly familiar with it. A similar instructive tool, in the form of a manual over two hundred pages long, was produced for Parks Canada's Atlantic region in response to a general lack of understanding among park staff (Department of Indian and Northern Affairs, 1977).

These extensive efforts at educating users of inter-disciplinary surveys suggest that this approach ranks low in terms of communicability. Contributing to this handicap is the fact that, from only one map, users cannot consider resources separately for their specific land use implications. By lumping the original data together into a single map unit, information, potentially useful for planning and management purposes, becomes inaccessible to the user; in effect, the data are lost.

Over the past ten years all of Parks Canada's five regions have adopted the inter-disciplinary approach. It was seen as the theme tying together Parks Canada's entire Natural Resource Program. In a critical review of this program, Cowan (1977) expressed his apprehension about this move:

> The hope has been expressed that the new approach will produce more information more quickly, less expensively, and in more easily useable form . . . Whether or not it can live up to the hopes of its proponents remains to be seen. In my view the biophysical approach should have been more thoroughly tested for its capacity to improve the quality of management of our National Parks before its adoption as the overriding theme. The sacrifice in depth, specificity and range is not inconsiderable and there may well have been better alternatives for certain Parks or for sectors of the data for all Parks. (Cowan, 1977, p.27)

To the author's knowledge, the Banff/Jasper project represents, to date, the most refined inter-disciplinary survey used by Parks Canada; yet during its preliminary stages, Cowan had serious reservations about its future applicability. He suggested that:

> . . . the entire biophysical process with all its extensions and elaborations [be tested] on a smaller scale - as for instance Waterton Park or one of similar size - and . . . the results evaluated before embarking on so massive and expensive a task as Banff, Jasper . . . (Cowan, 1977, p.29)

Cowan further suggested that the entire park, especially its wilderness areas, need not be surveyed to the high level of detail (1:50,000) specified in the terms of reference for the project, but rather that this apply primarily to areas in which significant human activity has occurred or is likely to occur (1977). Four years later, as the data came in and the results were being applied, it appears that Cowan's reservations were justified. For example, results from the wildlife component of the survey have found their greatest applicability in localized areas associated with existing or potential human development: for the location of campgrounds, picnic areas and other facilities; for the realignment of the 1A Highway in Banff; and for the analysis of mountain goat habitat in the vicinity of the Lake Louise ski area (Holroyd, 1980a). For broader land use planning decisions for wilderness areas, the survey results have found little application.

Though the inter-disciplinary approach by no means has been abandoned by Parks Canada, a recent manual from their Natural Resource division proposes a fundamental change to the established methodology:

> . . . when data are originally presented in an ecologically-integrated or biophysical framework [they] must be subsequently "disintegrated" to portray the distribution of specific resource components . . . For example, if soils or tree cover maps are likely to be required for a park where the Basic Resource Inventory utilized the "integrated approach", then considerable "analysis" will be required if component maps are desired (maintenance of unit boundaries is essential

> in such a case). In addition, information used by visitors, students and others with a less technical interest is frequently best presented in its simplest form. (Parks Canada, 1980, p.27)

This trend towards *disintegration* of resource data reflects what may be a general discontent among users with the "one convenient package" (Environment Canada, 1980) displaying so-called ecologically-significant land units. Failure to derive practical meaning from inter-disciplinary surveys because of the loss of resource-specific data no doubt has greatly restricted potential applications of this approach.

CHAPTER 3

RESOURCE SURVEY DESIGN

A HYBRID RESOURCE SURVEY APPROACH: THE ABC METHOD

Emerging from the foregoing appraisal of resource survey types are two opposed viewpoints which are of fundamental importance to the design and ultimate utility of resource surveys. Proponents of the inter-disciplinary approach assume that the ecological foundation of land use planning is strengthened by viewing different resource categories for what they are - *inseparable* components of the land (Nelson et al., 1978). Antithetical to this concept is the thematic view of the land and its resources, a view best illustrated by non-evaluative multi-disciplinary surveys. Bailey et al. strongly favour this latter perspective:

> Resources are usually too different and their interactions with other resources are too complex to be combined in the same classification. Each should be *classified separately* by its own intrinsic characters and in the context of values and uses for society. Once classified independently, different resources can then be compared objectively to study their interactions (emphasis added). (Bailey et al., 1978, p.652)

The trend towards "disintegration" among inter-disciplinary survey users suggests that Bailey's viewpoint may be more practical for ESA planning purposes. However this trend, carried too far, could lead to excessive fragmentation of resource categories, resulting in confusion among potential users as shown in Chapter 2.

The resource survey approach described here attempts to find a balance between these two extremes by promoting an optimal level of data integration. To overcome the problems created by resource data which are either too integrated or too fragmented, a hybrid multi/inter-disciplinary survey approach was designed which employs three levels of mapping and postpones the integration of data from different resource categories until after they have been independently analyzed with respect to land use implications. As such, the advantages to users are that land use opportunities and constraints arising from specific resource categories can be better appreciated and more effectively acted upon by respective decision makers than if all data were pooled from the outset (as in strict interdisciplinary surveys). As well, by retaining the data in their integrated *and* non-integrated forms, important ecological relationships both within and between resources can be elucidated.

In accordance with the principles outlined in Chapter 1, the following approach provides a methodological framework for the analysis of abiotic, biotic and cultural (ABC) data in a balanced and complementary fashion (after the work of Dorney, 1976).

Figure 3 shows the main elements of an ABC resource survey approach in the form of a conceptual model which is divided vertically into three stages representing increasing levels of data integration. The boxes represent the idealized map product at each level and correspond, horizontally, to the abiotic, biotic and cultural resource categories. In the remainder of this chapter, the main concepts and principles upon which an ABC resource survey should be based are explained first. Then specific procedures for the biotic component of this approach are presented. Examples of map output illustrating the major stages of the resource survey are included at the end of this chapter.

Level I

At Level I the approach can be recognized as non-evaluative multidisciplinary in that raw data from many resource variables are presented on maps of similar scale and format. The many types of variables required to meet information requirements for ESA planning can be accommodated at this stage. As well, the most can be made of pre-existing information when available. According to the model, data for each of the ABC components are presented on two sets of maps, one displaying structural type data and the other, functional. The former set depicts a structurally-oriented view of the environment, describing selected features of a given ESA. The view depicted by the latter map is more functionally-oriented, delineating areas within which major ecological processes occur. The types of data which are included on Level I maps and the means for their acquisition are presented below. Implicit in the following discussion is the requirement that all of the information is collected and analyzed by a logistically integrated team.

Within the abiotic component, structural variables could include properties and conditions of abiotic materials (texture, composition, moisture regime, temperature regime) and morphology (slope, relief, association of landforms, drainage patterns). Functional variables include processes associated with the origin (morainal, alluvial, lacustrine, marine, colluvial, etc.) and contemporary modification (solifluction, mass wasting, nivation, erosion, thermokarst action, avalanches, etc.) of abiotic materials. Stereoscopic interpretation of intermediate scale black and white aerial photographs is the primary mode of acquiring and mapping such information. LANDSAT satellite imagery is a more appropriate tool for deriving regional types of information such as physiographic limits, the

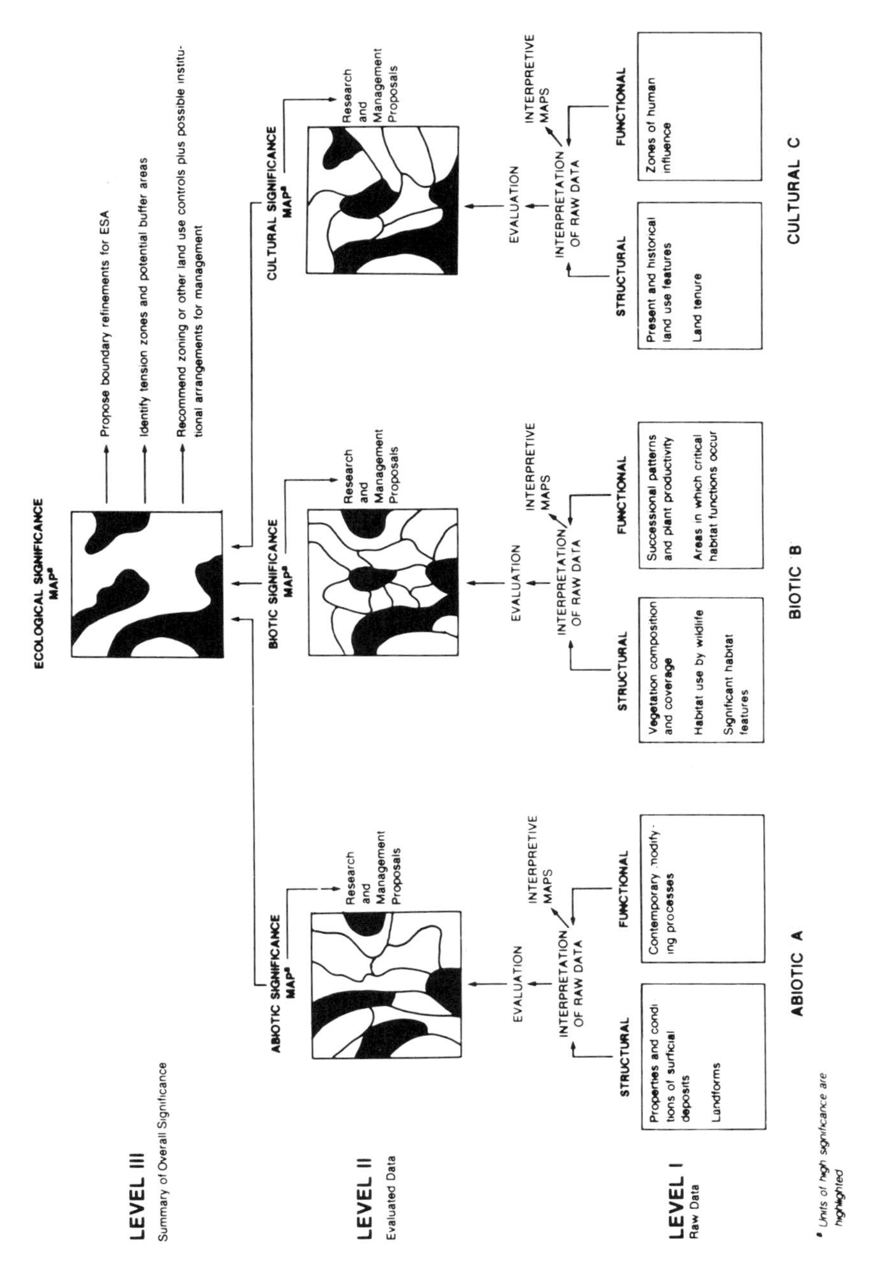

Figure 3 ABC Resource Survey Approach for Environmentally Significant Areas

definition of major drainage patterns and the limits of ice, snow or tree cover.

Within the biotic component, structural variables related to existing vegetation communities include physiognomy, species composition and coverage, age and fire history, while those related to wildlife are indicators of habitat use and some measure of habitat suitability for particular species. The functional map displays data on special or unique habitats within which important wildlife habitat functions occur including reproduction, feeding, migration and/or winter protection. This map may also include important information on successional patterns and concentrated plant productivity. For the structural information, LANDSAT imagery is used to identify broad vegetation patterns and to delineate representative areas for sampling in the field; functional information is derived from intermediate scale black and white photos, topographic maps, local interviews, field checks and a review of relevant literature.

Within the cultural component, structural variables include archaeological sites (pre-caucasian grave sites, burial grounds, tool caches or campsites), historic sites (old buildings, cemetaries, historic trails or water routes), aesthetic resources (view ratings, landscape personality, scenic roads, areas of local symbolic value), land tenure and transportation routes (road, rail, air, water). Functional variables relate to zones of human influence and activity (hunting/trapping/fishing areas, present and historical land use patterns). Interviews and literature searches are the prime sources of data for most of the cultural information; air photo interpretation and field investigations on past or present land use supplement these data. Cultural data are mapped using a parametric approach much like Lewis' (1964, 1965) in which "nodes of interest" and "environmental corridors" are recognized as the data are acquired.

In proceeding to Level II, all of the Level I maps are retained to avoid loss or masking of the baseline raw data.

Level II

In order to proceed to Level II, raw data are translated into a more meaningful form by employing interpretive indices which enable the comparison of natural and cultural resource values or constraints within an ESA. Sets of indices are derived from a mixture of ecological theory (e.g. uniqueness), human values (e.g. historic importance) and practical management considerations (e.g. vulnerability). The selection of appropriate interpretive indices depends upon such factors as: (a) the resource component being analysed (A, B or C); (b) specific user objectives; (c) ecological characteristics of the study area; and (d) the level of detail of raw data. A critical review of possible indices which are currently popular is presented by Margules and Usher (1981). Following interpretation, the structural and functional data within each of the three resource components are overlain and cartographically integrated. This procedure yields the Level II map units which are then evaluated so as to identify units which deserve highest priority in land use planning considerations, i.e. those units of highest *significance.* As such, Level II involves the systematic ranking of map units in terms of their relative significance - high, medium or low - with respect to each of the ABC categories.

Level II of the model can be recognized as evaluative multi-disciplinary in that resource information is evaluated within the context of distinct resource categories. Characteristic of this method is the low number of categories (in this case three) which allows for easy communicability to both technical and lay audiences. As well, replicability of results is high, given a set of precisely defined evaluation criteria for which assumptions are made clear. A unique advantage of Level II is that it allows independent analysis of the abiotic, biotic and cultural systems.

Once units of highest significance have been identified, their resource attributes are described in detail, specific management recommendations are made and research needs and opportunities related to these attributes are identified. In a hypothetical example, a map unit from a Level II biotic map is ranked high in significance due to its importance as habitat for a localized mountain sheep population. Research needs might include summer and winter aerial surveys of the identified habitat in order to gather population data on the sheep and to determine their movement patterns within the unit or between it and other units. Immediate management recommendations could involve a temporary suspension of hunting privileges in the area to ensure interim protection of the sheep population until adequate data become available for use in drawing up more specific management plans.

From this example, it can be seen that uni-disciplinary surveys may be suitable for more intensive, larger scale studies on high priority units identified in Level II. In some cases, where a specific development project is proposed, say a cottage subdivision or a mine, some form of multi-disciplinary survey may be more appropriate in providing the necessary scope of data required for site planning. In either case, the approach provides great flexibility in that the required type and level of detail of resource data for third phase studies may be acquired and analyzed using the most appropriate type of survey method; "data overkill" would be avoided thereby assuring cost-effectiveness and relevancy of survey results.

To proceed from Level II to Level III, map units of high, medium or low significance within each of the ABC categories are first assigned a rank of 3, 2 or 1 respectively according to the pre-determined criteria. Then the three Level II maps are overlain and their ranked units are numerically integrated. This step involves summing of the ABC ranks through cartographic overlap, drawing unit boundaries around areas with a similar sum, and then ranking the resultant units. The outcome of this procedure is the Level III map which displays units whose ranks (again, high, medium or low) correspond to their relative ecological significance.

Level III

At Level III, the model shares some characteristics of the inter-disciplinary approach in that:

1. The final map product represents the outcome of logistically integrated teamwork among several resource disciplines.
2. Diverse ecological data are integrated, then conveniently nested into one set of map units.

However, unlike conventional surveys of this kind, the data are in evaluated, rather than raw form when they are integrated, or, in short, ecological integration is the last rather than the first step in data manipulation. The advantages of adopting this strategy are discussed further below.

The primary function of the Level III map is to focus the user's attention upon particular areas in which ecological values are concentrated. The objectives of Level III are threefold: first, to show the spatial distribution of ecological significance within an ESA; second, to display, on one map, all of the sites which deserve high priority in land use planning and management decisions; and third, to display the degree of geographic overlap of these sites so that relationships among abiotic, biotic and cultural resource components can be determined. By studying the degree of geographic overlap, spatial coincidence and other relationships among mapped information, the nature of management problems and possibilities can be ascertained and research needs more precisely identified.

To clarify the latter objective, another hypothetical example is given. On the Level III map, the boundaries of a high ranking unit correspond roughly to a broad, steep-sided mountain valley. Boundaries are defined by the mutual overlap of three Level II map units, all of which were independently evaluated and deemed highly significant for various reasons pertinent to each of the ABC resource categories: abiotically, because it includes a unique assemblage of geological features and an area of high avalanche potential, biotically because it includes unusually diverse plant communities and an area important as winter range for moose and culturally, because throughout the valley is an abundance of archaeological and historic sites.

In this hypothetical case, a concentration of significant abiotic, biotic and cultural features is located within a relatively discrete valley ecosystem. Presumably, the high degree of overlap revealed in defining boundaries at Level III indicates that, within that unit, a high degree of ecological inter-relatedness exists among the resource variables displayed on the Level I map. In contrast, Level III map units revealing little or no geographic overlap among Level II map units would indicate lesser degrees of inter-relatedness. Thus, an important advantage of Level III mapping is that, through the explicit display of coincidences and discrepancies between abiotic, biotic and cultural boundaries, the degree of inter-relatedness among selected ecological phenomena can be revealed. Properly applied, the Level III map would be used in conjunction with both the Level II and Level I maps to ascertain the nature of ecological relationships in specific sites and, on the basis of this knowledge, to identify ecological constraints and opportunities for particular land uses within those sites. And so, collectively, the maps of the proposed model would prove a sound basis for truly ecological planning.

Other applications of the Level III results for prescriptive purposes include:

1. The proposal of boundary refinements for ESAs.
2. Recommendations related to buffer zones.
3. The formulation of zoning policies or other land use controls for units of high ecological significance.

As an alternative or supplement to traditional methods of delineating park or reserve boundaries - through recognition of physiographic or jurisdictional divisions - boundaries for ESAs could be systematically refined by proposing the inclusion of peripheral units of high ecological significance, particularly those units which encompass resource values that justified the original ESA designation.

Related to boundary delineation is the subject of buffer areas. Once a boundary has been proposed, units of high ecological significance which are close to but outside of this boundary could be highlighted as possible "tension zones" - units outside ESAs in which inappropriate land use could pose a threat to the protection or preservation of ESA resource values. The position and resource attributes of these units could guide recommendations concerning the location of buffer areas and the type of land use controls within them that would be appropriate.

The foremost question to be addressed in regard to zoning or other land use controls in ESAs is how should ecological information be applied towards the conservation of resource values within units of high ecological significance? In more general terms, Grigoriew describes two ways in which such information can be applied to land use planning.

> 1) . . . to influence the direction and change in land use, the choice of sites for various activities and developments, and the intensity of use of a site in order to minimize the disruption or destruction of natural ecosystems . . .
> or
> 2) . . . [to identify] . . . particular sensitive areas from which certain uses should be excluded for the perpetuation of certain values associated with those particular areas. (Grigoriew, 1980, p.9)

Since the main thrust of the proposed approach is to translate ecological information into the identification of priority areas for planning, land use controls in ESAs should be based primarily on the latter application. In short, any zoning, regulation or permit system for ESAs should emphasize the ability of particular sites to sustain certain land uses rather than to match sites with demands.

PROCEDURES FOR THE BIOTIC COMPONENT OF AN ABC RESOURCE SURVEY

This section describes a set of procedures for the acquisition, interpretation, evaluation and presentation of biological data for ESA planning purposes. Specific procedures are presented in sufficient detail to allow their:

1. Immediate application to ESAs in a standardized fashion.
2. Logistical integration with abiotic and cultural procedures.
3. Refinement or modification as is judged necessary in future.

For Level I structural variables, a landscape approach to classification was adopted in which distinctive areal units showing similar patterns of vegetation are recognized and mapped through remote sensing. Collectively, these units, referred to as *Biotic Land Units* (BLU), provide a geographic framework with which to describe existing vegetation communities within an ESA, as well as to identify faunal assemblages and significant habitat features which characterize these communities.

In choosing vegetation as the prime classification criterion for biotic land units, an important ecological relationship between vegetation and wildlife was assumed, viz. for most wildlife, the composition and physiognomy of plant communities are the major factors determining habitat suitability. It should be noted that although this ecological relationship has been validated by several authors (DeVos and Mosby, 1971; Thomas, 1979), biotic land units are not meant to represent discrete ecosystems.[13] Rather, they are no more than conveniently recognized units into which biological data on structural variables can be meaningfully incorporated or extrapolated. Also, they form a framework for field sampling to assure coverage of major communities throughout the area for subsequent interpretation of significance for planning. For referencing purposes, biotic land units are grouped into several "Landmark Areas" which are bounded by major drainage systems, mountain ridges or roads and which are named after prominent landscape features within their boundaries.

For Level I functional variables, a parametric approach to classification was adopted in which *Special Habitat Zones* are identified. These zones provide habitat functions critical to the survival of wildlife, particularly large mammals, raptors and waterfowl. Among these functions are seasonal movement, winter protection and foraging and rearing of young. Also included in these zones are areas which provide a variety of habitat functions to a relatively high diversity of species, namely riparian corridors and areas characterized by cliffs, talus and caves such as canyons or scarp faces.

----

[13] Rowe (1961) has shown that plant communities in themselves should not be equated with ecosystems.

The approach is parametric in that special habitat zones are identified by recognizing, primarily through remote sensing, a complex of environmental attributes which are assumed to be relevant to wildlife. These attributes may or may not include vegetation and such abiotic factors as topography, elevation, substrate, slope, aspect and microclimate. In certain cases, such as rearing areas and movement patterns, the only parameter to identify a special habitat zone is the reported presence of animals routinely performing a particular function in a particular area.

In all cases, special habitat zones should not be regarded as relatively distinct geographic entities like Biotic Land Units; rather, as the term "zone" suggests, they should be regarded as generalized areas, due to the variablity both in climate and in behaviour of large mammals, and to the arbitrary nature of human-imposed boundaries on complex animal habitats.

The functional map may also include data on areas of concentrated plant productivity, unique or representative successionary patterns or other information related to important functional aspects of vegetation.

Data Acquisition

Procedures for data acquisition are explained below in the context of three sections - *Prefield, Field* and *Postfield* - according to the sequence in which they should be undertaken. There is usually some overlap between groups; for instance, interviews may extend well into the field season. However, the schedule is adhered to as closely as possible, completing each set of procedures before proceeding to the next, to ensure systematic results and an optimal use of time.

Though untested outside of the Yukon ESAs, the Biotic Land Unit concept could be applied elsewhere to natural areas of a different scale or relative biological diversity (generally greater diversity in warmer climates). At larger scales for which alternative remote sensing tools as described below are required, similar mapping criteria could be used to delineate broad patterns of vegetation. In all cases, the use of infra-red enhanced images is recommended in order to facilitate the mapping process.

Prefield Procedures

(a) *Pre-typing of Biotic Land Units:*

This step involves the recognition and mapping of Biotic Land Units from LANDSAT satellite imagery. LANDSAT emphasizing infra-red reflectance was chosen as the most appropriate remote sensing tool for several reasons. First, infra-red LANDSAT images allow a small scale regional overview of vegetation patterns while providing greater contrast between plant communities than black and white photos of much larger scales.[14] Second for most of northern Canada - the focus of our ESA research - the only other remote sensing tools available are outdated black and white photos (1:63,000) which are unsuitable for mapping present vegetation. Third, for reconnaissance level mapping of vegetation, LANDSAT has elsewhere been proposed as a useful tool for both lowland (Thie, 1974b) and alpine areas (Jaques, 1976) environments. Fourth, LANDSAT images can be photographically or optically enlarged (from 1:1,000,000) to a convenient scale of 1:125,000 without loss of resolution quality. And last, LANDSAT images are available in several formats including various types of prints or transparencies.

LANDSAT interpretation depends on the differentiation of areas based on their spectral signatures. These signatures correspond on the ground to recurring patterns of vegetation and are recognized by observable differences in the colour, texture and tone of the image. Up to ten predominant signature types or "themes" can be accurately recognized through manual interpretation, forming the basis for preliminary delineation and classification of map units.[15] By using a zoom transfer scope and a 20 x 20 cm LANDSAT positive image, themes or classes can be recognized and simultaneously translated into map units drawn onto a 1:125,000 base map of the study area. As a general rule, the minimal size for a map unit should be 2.5 cm in diameter if rounded and not much less than 1.0 cm wide if the shape is long and narrow. This means that in the field, the minimal area to be considered should be not less than 6 km^2.

The resultant map units should be coded according to theme by a system of arbitrary symbols or colours. At this point, anomalous units which are difficult to link with a predominant theme should be highlighted for later inspection in the field. In some cases, temporal or other constraints may not permit the whole study area to be pre-typed; in this instance, at least enough interpretation of the image must be done so that the entire diversity of themes is represented in the delineated map units. The boundaries for map units should be subsequently refined through the use of LANDSAT images from different seasons,[16] available black and white photos and through experience in the field as LANDSAT themes become translated into recognizable plant communities or community complexes.

----

[14] Thie (1974b) cites studies indicating that 1:120,000 scale colour infrared imagery can provide the equivalent amount of information as 1:60,000 black and white photos.

[15] The possibilities and cost effectiveness of substituting computer-assisted mapping procedures for manual interpretation is a topic of ongoing research at the University of Waterloo.

[16] The time of year influences the spectral signature of plant communities. Seasonal imaging may assist the differentiation of themes especially those that correspond to conifer associations where changes in the condition of substrata have a marked effect on reflectance.

(b) *Identification of Special Habitat Zones:*

Since there can be several types of special habitat zones, they are identified and mapped by following different procedures. The importance for wildlife of each type of zone, along with the way in which they are recognized, are presented below. Comments concerning the importance to wildlife of riparian zones and cliff/talus/cave complexes are distilled largely from Thomas (1979).

Riparian-zones are areas which include, or are in close proximity to, aquatic systems and which are characterized by the presence of vegetation that requires free or unbound water or conditions high in moisture relative to the total area of study (Thomas, 1979). These zones can vary considerably in size and vegetative cover but all are of special importance to wildlife for the following reasons:

1. For many types of wildlife, riparian zones may offer, in one place, three critical habitat components - food, cover and water.
2. Local edaphic factors often lead to increased diversity in the species and structure of plant communities and thus in the diversity of available habitats.
3. The linear shape of many riparian zones maximizes the development of edge which is so productive of wildlife.
4. For migrating or dispersing wildlife, riparian zones may serve as connectors between habitats.
5. Microclimatic characteristics which are common to most lowland riparian zones and attractive to wildlife include increased humidity, shade and air movement, as well as an abundance of thermal cover.[17]

Though riparian zones may include both lentic (standing water) and lotic (running water) habitats, only the latter are considered here. From a planning perspective, it would be impractical to designate the shores of all lakes and ponds as special habitat zones; variation in habitat importance would be too great to justify such a generalization.

Black and white photos provide the most accurate means for delineating riparian zones. Rivers and major streams as well as marked contrasts between the surrounding vegetation are sufficiently evident to allow these zones to be broadly depicted on a 1:125,000 base map. Topographic maps can aid in this procedure by suggesting moist areas of relatively uniform drainage. The use of LANDSAT images together with topographic maps is a feasible though less accurate alternative.

----

[17] In many subalpine valleys, riparian zones may exhibit the "reversed treeline" effect. Though some of the above microclimatic conditions may be absent in these situations, the development of edge contributes to this zone's biotic richness.

Only water courses over approximately 6 kilometres in length should be considered for mapping.

Cliff/talus/cave-complexes are rather special habitats in that they:

1. Occur infrequently.
2. Are comprised largely of geomorphic features.
3. Often contain species that cannot live elsewhere.
4. Cannot be artificially created or manipulated for management purposes.

These complexes provide relatively secure habitats for the reproduction and feeding of many species of wildlife.

Cliffs are steep, vertical or overhanging rock faces that provide physical protection for wildlife and concentrate a variety of birds and mammals[18] into relatively small but stable environments. Raptors such as the golden eagle, peregrine falcon and American kestrel are particularly dependent on cliffs, as well as such passerines as Say's phoebe, cliff swallow, Townsend's solitaire and gray-crowned rosy finch. Large mammals using cliffs for reproduction, feeding and often for escape terrain from predators include mountain goat and Dall sheep. Cliffs are also important for bats, wolverines, long-tailed weasels and ground squirrels.

Talus is the accumulation of broken rocks at the base of cliffs or other steep slopes. Species which depend on this habitat for reproduction and hibernation functions include the hoary marmot, white-tailed ptarmigan and pika. Talus usually supports an edge of herbaceous or shrubby vegetation providing a food supply for these species.

Caves are natural underground chambers that are open to the surface. They provide shelter from extreme weather conditions, from varying degrees of darkness (for hibernation) and physical protection from predators. For the following animals, caves are particularly important as sites for reproduction: wolverine, bat, marten, fisher and porcupine, as well as canids and weasels.

Since these physical habitats - cliffs, talus, and caves - often occur together naturally, they have been combined into one type of habitat zone. Cliff/talus/cave complexes are most accurately identified and mapped through stereo interpretation of black and white photos. These complexes can be recognized as areas of abrupt and jagged relief, with little or no apparent vegetative cover, and should be delineated as special habitat zones on the 1:125,000 base map. As with riparian zones, LANDSAT and suitable topographic maps provide a second choice remote sensing package if black and white photos are not available. As well, like riparian zones, cliff/talus/cave complexes are typically linear in shape so the 6 kilometre minimal length limit should guide in the identification of these zones.

----

[18] Example wildlife species are primarily from northern Canada.

Snow-free-zones are gently sloping, usually south-facing alpine areas which, due to exposure and local air flow patterns, are free of snow for all or part of the winter. Where suitable vegetation is present, these zones may serve as critically important feeding areas for ungulates, especially Dall sheep and caribou. Identification of these zones involves the use of "quicklook" LANDSAT imagery.[19] This system provides a multi-date record of snow conditions allowing researchers to locate vegetated areas which are free of snow all winter or from which snow recedes. Inspection of conventional LANDSAT imagery can aid in the delineation of these zones with respect to the degree of vegetative cover. A zoom scanner is used to transfer this information to the base map.

Staging-areas are large areas of open water which are available to massing waterfowl during their spring migration northward. These are identified by scanning "quicklook" imagery for lake or river surfaces free of ice in the late winter.

Information concerning the next four types of special habitat zones - winter range, rearing grounds, migration routes and spawning areas - is acquired primarily through literature research and personal interviews, the procedures for which are expanded upon in section (d) below.

The term *winter range* is distinguished from "snow-free zone" in that the former term refers to habitat used by a specific species rather than possibly several as in the latter, though there may be some geographic overlap. Winter range characteristically provides both feeding areas and thermal cover. The reported presence of a viable ungulate population occupying a well-defined area over several winters is a pre-requisite for the identification of winter range.

Rearing grounds are small, ephemerally used, but extremely critical sites where large mammals are born and cared for during their first few days or weeks. They include moose and caribou calving grounds, Dall sheep lambing areas, mountain goat kidding areas or denning sites for wolves, foxes, coyotes and bears. Empirical evidence should exist which indicates routine use of a given site by these species for that site to be considered for mapping.

Migration-routes are routes of travel used routinely by wildlife, primarily ungulates, in their seasonal movement from one habitat to another (Thomas, 1979). These routes may include traditionally used watercrossings, forested corridors or more generalized movement areas. Depending on the accuracy and detail of available data, migration routes may be shown on the base map as arrows depicting generalized movement patterns or as a bounded zone encompassing a specific area of critical importance.

----

[19] See Lavigne et al. (1977) for other wildlife-related applications of "quicklook" imagery.

(c) *Field Preparation:*

Most of the field preparation revolves around how, where, when and by whom should the data describing the pre-typed biotic land units and special habitat zones be collected. How the data are collected is a matter of planning for transportation both to and within the field. Air support will be necessary for most ESAs in frontier areas. Where to sample is of crucial importance, because at stake is the optimal use of time and money, as well as the validity of results. Biotic land units to be sampled should be chosen on the basis of several factors, given here in order of priority. Sampled units should include:

1. At least four to five units for all of the 10 (+) LANDSAT theme types so as to investigate the consistency of vegetation within each theme.
2. As many anomalous units (units not corresponding to a predominant theme) as possible, especially the larger ones.
3. Units in which development exists or is likely to occur. Attention should be focussed in valley areas where roads and human settlement are typically concentrated.
4. Units which correspond to areas sampled by the members of the team dealing with the abiotic and cultural components.
5. Units in close proximity to each other and/or conveniently accessible in order to minimize field time and expenses.

The selection of special habitat zones to be sampled should be guided by points 3, 4 and 5 above. As well, the sampling of each type of special habitat zone within the ESA should be a prime consideration (providing the season is appropriate).

Careful scheduling of work activities is essential in order to meet temporal or financial constraints. A systematic though flexible critical path for data acquisition is useful, based upon such factors as blooming times for alpine vegetation, nesting dates for birds, rearing periods for large mammals, high water times and hunting seasons.

(d) *Literature Search and Interviews:*

The objectives of this step involve much more than initial familiarization of the field crew with the study area. Most importantly, this step aims to fill information gaps which cannot be filled in a season's field work. This information includes: the identification of winter ranges, migration routes, rearing grounds, previous wildlife sightings, local migration patterns of waterfowl, rare or unusual plant species or communities, rare or unusual animals, historical trends or events related to animal populations and plant communities and proposed or existing areas of development. Another objective of this step is to identify potential sites of special concern to planning. Lastly, this step serves as an inventory of the existing level of information on the study area. This knowledge should be especially valuable when recommending research needs for phase three studies in land units of high biotic significance.

Government agencies, consultant's reports and bibliographic maps are the primary sources for this information.[20] While in the field, all attempts should be made to interview local residents; in particular, hunters, trappers and outfitters should be contacted. Not only are many of these people knowledgeable about the local biota, but also about unmapped access routes and trails.

All information obtained through literature research and interviews should be recorded in field books and/or annotated onto a working map for later compilation.

Field Procedures

a) *Preliminary Overview of ESA*

Before sampling, it is advantageous for the field crew to gain a general perspective of the ESA to be surveyed. ESAs of small size or with good access can be toured by vehicle or by foot. An overview in larger, less accessible ESAs may require air support. Though only a few days should be necessary for this step, the time will have been well spent since the preliminary overview provides an opportunity to explore the problems and potential of access, to evaluate and refine the field sampling plan and to begin to match the LANDSAT themes and pre-typed land units to vegetation patterns in the field.

b) *Field Sampling*

To clarify field sampling procedures related to vegetation, three relevant terms are defined below:

Physiognomic Class - category describing the general appearance of vegetation with respect to form, structure and/or dominant plant cover. Examples could include: open conifer forest, closed ground shrubland, conifer swamp or partially stabilized sand dunes (see below for complete list).

Vegetation Type - descriptor referring to the plant species which comprise the greatest amount of per cent cover within a particular area. For each type of physiognomic class represented in a study area, there may be more than one vegetation type. As a hypothetical example, vegetation types for the closed conifer forest category could include: (a) white spruce/feathermoss/forest lichens;

----

20 Pertinent agencies and references for the Yukon only appear in Appendix C.

and (b) lodgepole pine/willow/shrub/soapberry/kinnick.

Vegetation Community - distinct plant assemblage to which one physiognomic class and one vegetation type are assigned. For example, here are two hypothetical vegetation communities: (a) a mixed forest dominated by trembling aspen/white spruce/willow shrubs/grass; and (b) a shrub bog dominated by willow shrubs/wetland mosses/sedges.

Major/Minor Vegetation Community - LANDSAT themes - and thus biotic land units - usually contain more than one vegetation community. The community with the greatest area coverage within a biotic land unit is referred to as the major vegetation community; other communities with proportionately less coverage are referred to as minor vegetation communities.

Using aerial photographs or LANDSAT, prospective sampling areas can be selected within the major vegetation community of previously designated BLUs. Such factors as homogeneity of tone, texture and colour (if LANDSAT), as well as ease of access should be considered at this point. In the field, community representativeness and homogeneity of environmental conditions (slope, landform, soil texture, drainage, species composition, physiognomy and coverage) should guide the final selection of the sampling area. In some areas a complex of two easily distinguishable vegetation types may typify the major community of a biotic land unit. Such bipartite vegetation communities are not uncommon, particularly in alpine environments where regular variations in microtopography strongly influence vegetation characteristics. In such cases, both types of vegetation should be sampled to provide a complete description. Care should be taken to sample in areas free from such localized human disturbances[21] as roads, trails and buildings as these may directly or indirectly (through microclimatic effects) influence vegetation and/or wildlife characteristics.

To describe the biotic variation adequately within one community, a wandering transect approach[22] was adopted for field sampling. This technique requires that two field persons walk separately through a sampling area until each is confident that he/she has observed the vegetation and wildlife characteristics which typify the community under study. At this point, two fieldsheets (Appendix D), as described and explained in Appendix E, are independently filled out, the results of which are averaged during subsequent data analysis. Other information related to minor communities (vegetation communities of a comparable

----

21 As distinguished from generalized human disturbances such as logging or grazing which may characterize a community.

22 As opposed to a fixed plot approach (e.g. see Walmsley et al., 1980) which, in a pilot survey, was found to provide inadequate coverage.

but lesser areal coverage within BLU than the major community), significant habitat features and human-related influences may be added to the fieldsheets as field photographs are inspected and literature and interview results are compiled.

The fieldsheet presented in Appendix D has two objectives:

1. To describe the outstanding biotic characteristics associated with the sampling area in the major community.
2. To provide a summary description of the range of biotic resources within the entire BLU.

To meet both objectives the fieldsheet has been designed to capture readily observable features so that, with little training, a field team of two may systematically and rapidly collect "diagnostic" information. To aid computer storage and compilation of the data, the fieldsheet information is recorded in checklist or coded form. A detailed description of field procedures which follows the format of the fieldsheet is presented in Appendix E as a manual to be used in the field; further guidelines regarding site description and estimating foliage cover visually are given in Appendices F and G respectively.

DATA INTERPRETATION

From the outset, an important precept in designing the survey was that, within any ESA, some biotic units are more deserving of protection/preservation measures than others. To identify these units, the presentation of raw biological data in mapped form is not enough. Beyond showing the distribution of and spatial relationships between biotic resources, such maps convey little information to aid the establishment of planning priorities. Therefore, the next step is to translate the raw data for each unit into a more meaningful form through a system of interpretive indices pertinent to particular resource values. Only after this interpretation has been completed can biotic land units be evaluated and those that have greater significance for planning be identified.

The practise of applying interpretive indices to the evaluation of natural areas is well established (Gehlbach, 1975; Goldsmith 1975; Sargent and Brandes, 1976; Sauchyn and Bastedo, 1980; Wright, 1977; Hrabi et al., 1979; Van der Ploeg and Vlijm, 1978). However, in spite of the practical advantages of interpreting data using indices, the inherent limitation of the process should be stressed. In any attempt to devise indices, "a hoard of assumptions must be made" (Inhaber, 1976, p.30). Depending upon the validity of those assumptions, quite different results could emerge during the evaluation stage; faulty, inappropriate or ill-defined assumptions could have serious consequences for the identification of significant areas.

From these considerations the following guidelines were developed to be used in devising the interpretive indices for biotic land units. First the indices should be pertinent to values which are justified by existing biological knowledge and/or the best professional judgment available; second, they should be appropriate to the level of detail of the raw data; third, they should be based primarily on quantified data to minimize bias during interpretation; fourth, they should incorporate inclusive, non-overlapping criteria; and last, they should be precisely defined and explained.

Based on quantitative relationships among raw data variables, criteria are established for scoring each biotic land unit with respect to a set of interpretive indices. For each index, a score range of 1 to 5 is established to ensure that subtle differences among biotic land units can be detected.

It should be noted that in the following examples, criteria for all of the indices were established on the basis of actual raw data from an ESA in south-central Yukon (Bennett Lake/Carcross Dunes/Tagish Lake ESA). From these results, upper and lower limits meant to encompass even extreme cases were established for each index. Once a suitable range was developed, the intervening values were sub-divided into 5 criteria sets to accommodate a 1 to 5 scale.

Indices Related to Biotic Resource Values

Under this rubric, the following indices shown in Table 2 are defined and explained: (1) community diversity; (2) uniqueness; (3) recoverability; (4) faunal diversity; (5) faunal dependence; and (6) fire susceptibility. All but the first index apply specifically to a biotic land unit's major vegetation community rather than to the unit as a whole. Often this community may be comprised of a close association or complex of two or more distinct communities. In these cases, scores should be calculated separately for each community. Score values for particular indices should then be averaged to yield a final score for the unit.

Community Diversity

This index refers to the number of vegetation communities (major and minor) found within biotic land units; the more communities per unit, the more habitats are assumed to be available for various wildlife species. In order to be recognized as distinct, a community must be a minimum of one square kilometre in size.[23] Hence, for example, if a biotic land unit contained eight or more vegetation communities, the unit would be assigned a score of 5 according to the criteria presented in Table 2 under "community diversity".

----

[23] Below this size, vegetation communities cannot be resolved accurately from intermediate-scale aerial or field photographs.

INDICES	CRITERIA	SCORE
1. COMMUNITY DIVERSITY (number of vegetation communities > 1 km² within biotic land unit)	8 or more	5
	6 - 7	4
	4 - 5	3
	2 - 3	2
	1	1
2. UNIQUENESS (uniqueness quotient value related to major community)	1 - 9	5
	10 - 109	4
	110 - 209	3
	210 - 309	2
	310+	1
3. RECOVERABILITY (relevant characteristics of major community)	- Conifer forest at tree-line - 4 or more indicators of frost action - Canopy 200+ years old	5
	- Wetlands - Communities susceptible to erosion or slumping - 3 indicators of frost action - Canopy 150-200 years old	4
	- 2 indicators of frost action - Canopy 100-150 years old	3
	- 1 indicator of frost action - Canopy 50-100 years old	2
	- No indicator of frost action - Canopy 50 years old	1
4. FAUNAL DIVERSITY (number of species currently and/or potentially using major community's significant habitat features[b])	a) Mammals	
	10+	5
	8 - 9	4
	6 - 7	3
	4 - 5	2
	2 - 3	1
	b) Birds	
	29+	5
	23 - 28	4
	17 - 22	3
	11 - 16	2
	5 - 10	1
5. FAUNAL DEPENDENCE (% species currently and/or potentially occurring in major community for which versatility values are at or below median versatility value)	a) Mammals (median versatility value: 10)	
	50+	5
	45 - 50	4
	30 - 44	3
	15 - 29	2
	0 - 14	1
	b) Birds (median versatility value: 5)	
	50+	5
	45 - 50	4
	30 - 44	3
	15 - 29	2
	0 - 14	1
6. FIRE SUSCEPTIBILITY (relevant characteristics of major community)	- Closed lodgepole pine forest	5
	- Closed conifer forest other than lodgepole pine - Closed mixed forest with high birch and/or lodgepole pine component	4
	- Open conifer forest - Conifer woodland - Closed tall shrubland: willow - Closed low or ground shrubland: Labrador tea or crowberry - Mixed forest other than above - Deciduous forest	3
	- Graminoid meadow - Medium shrubland: dwarf birch and/or willow - Low shrublands other than above - Ground shrubland with scattered low and/or medium shrubs	2
	- Sparsely and non-vegetated areas - Wetlands - Ground shrubland other than above	1

[a] *Criteria for each index based on raw data from Bennett Lake/Carcross Dunes/Tagish Lake ESA, south-central Yukon Territory, Canada; extremes were established and intervening values subdivided to accommodate a 1 to 5 scale.*

[b] *Significant habitat features include: a) for mammals: favourable combinations of preferred breeding habitat, major food source, cover requirements, and/or ecotone, b) for birds: preferred breeding and/or foraging habitat.*

Table 2 Indices Related to Biotic Resource Values

Uniqueness

Both the frequency of occurrence and the relative areal extent of a vegetation community are indications of its relative uniqueness within a particular ESA. The equation below is based on the assumption that the largest of the least frequently occurring vegetation communities within an ESA would be most unique. Conversely, the smallest of the most frequently occurring vegetation communities would be least unique.

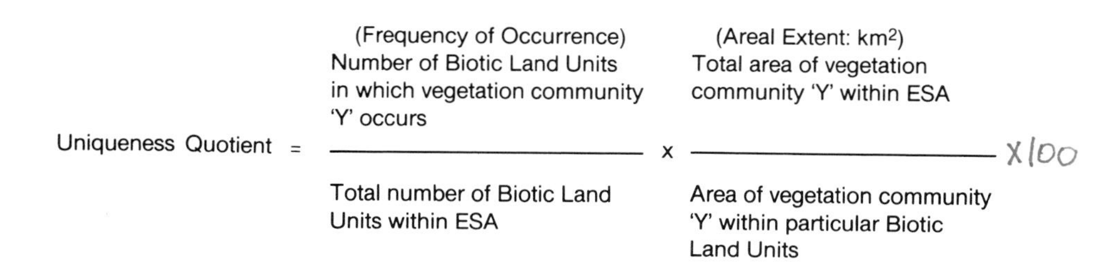

Quotient values are rounded off to eliminate decimals. The lowest value would be 1 corresponding to the highest uniqueness score.

Recoverability

Following severe disturbance of a vegetation community (intensive fire, logging, mining activities and so on) the existing vegetation may in some cases: (a) be replaced by different associations of plants (irreversible change) or (b) return only after extended periods of time (time lag) (Rykiel, 1979). Of concern then is the possibility of and/or the relative rate at which a vegetation community would return to its original composition following the disturbance, i.e. its recoverability.

Below treeline, the most convenient indicator of a community's recoverability is its age, older communities having a longer recovery rate. For scoring purposes, the age class of the dominant canopy strata is used to indicate recoverability in most lowland units.

Certain vegetation communities for which accurate aging is difficult or for which recoverability is influenced largely by local edaphic factors (microclimate, soil erodibility, etc.) should be given special consideration. Examples of communities in the Yukon include: (a) conifer forests at treeline, for which establishment may have occurred during a former climatic period which was slightly warmer than at present (Webber and Ives, 1978), thus reducing the possibility of its future recovery following disturbance; (b) wetlands, whose long-accumulated peats, susceptible to rapid decomposition if disturbed, indicate a community type with relatively low recoverability; (c) graminoid meadows on highly erodible glacio-lacustrine deposits; and (d) active or partially stabilized sand dunes upon which a thin veneer of vegetation plays a critical role in controlling the rate of eolian erosion and deposition.

Above treeline, recoverability of vegetation communities can be correlated with the intensity and character of frost action, the most important factor in sub-alpine or alpine community development (Sigafoos, 1952). The greater the influence of frost action on vegetation in these environments, the less successful will be a community in re-establishing itself following disturbance (Mackay, 1970). In this regard, the number of types of microrelief features indicating contemporary frost action (mud boils, stone circles, etc.) was assumed to reflect the degree to which vegetation is influenced by this action. (See Qualifying Remarks, p. 70).

Faunal Diversity

This index gives a measure of the relative diversity of wildlife species which may occur among the major communities of biotic land units. Mammalian and avian fauna in most cases are considered separately due to a greater number of bird species in relation to mammal species. For both groups, the number of species occurring in a given community is assumed to reflect faunal diversity. The resultant scores are recorded and averaged separately in order to convey the mammalian, avian and overall faunal diversity of a community.

Faunal Dependence

Faunal dependence indicates the degree to which wildlife species occurring in a given vegetation community are dependent on that community for the significant habitat features it provides. The calculation of this score involves three steps: (1) determine for each species, the number of vegetation communities in which it may occur, i.e. its *versatility value* (Thomas, 1979); (2) determine for each faunal group (birds and mammals) the median versatility value; and (3) determine the percentage of species occuring in a given community for which versatility values are at or below the median. As seen on Table 2 the higher this percentage is for a community, the higher would be its score, reflecting a high proportion of low versatility species. In many alpine communities, for example, most of the resident wildlife species are almost exclusively dependent on significant habitat features provided by these communities, conferring on these communities a high faunal dependence score.

Fire Susceptibility

The fire susceptibility index provides a predictive measure of the major vegetation community's potential to catch and carry fire in its present state. Based on both qualitative and quantitative data, the criteria for scoring presented in Table 2 are based primarily on variables selected from Novakowski (1970), Grigel et al. (1971), Kiil et al. (1973), Rowe (1972) and Hettinger et al. (1973), all of whom have created fire prediction systems of varying levels of detail. The system here presented is of a general nature only, derived largely from Hettinger et al. (1973, pp. 103-106). In creating this system, the following variables were considered: physiognomic class, vegetation type, type and per cent cover of organic litter, evidence of previous fire and fire periodicity. As a guiding rule, the greater the per cent cover in the uppermost strata of a community and the greater its coniferous and organic litter components, the higher would be its fire susceptibility score.

Indices Related to Human Influence

It should be noted that this part of the biotic survey is optional, to be undertaken only if the cultural component of the ABC survey approach is not undertaken concurrently. In such cases, raw data on "Human Related Processes Influencing Vegetation and/or Wildlife" (see section, Data Presentation on page 66 below) are collected and then interpreted using the indices below.

A set of indices for comparing the degree of human influence among biotic land units is devised using raw data variables associated with past, present, and potential land use activities (Table 3). The indices discussed below apply to the biotic land unit as a whole and include: (1) human impact; (2) existing development; (3) potential development; and (4) recreation activities. These indices are of a more general nature than those related to biotic resource values as indicated by the relative simplicity of calculations to determine score values.

An important assumption upon which all of the indices are based is that the number of land use activities provides a fair indication of human influence. However, for the indices, no consideration is given to the type or relative effect of these activities on biotic resources.[24] Though rather simplistic, similar indices devised by Hrabi et al. (1979) have proven useful in comparing natural areas with respect to human influence. As in their work, it was assumed here that the greater the number of land use activities in an area, the greater attention that area should receive in the establishment of planning priorities, hence the higher the score.

Human Impact

The more kinds of past land use activities evident in a biotic land unit, the higher the human impact score assigned to the unit. This index is useful both for identifying units which may require greater attention to protect biotic resources from further damage and in recognizing opportunities for monitoring long term impacts of particular activities on these resources.

----

[24] These and related considerations for ESAs are topics for more indepth research being conducted by other members of the ESA research team at the University of Waterloo.

INDICES	CRITERIA	SCORE
1. HUMAN IMPACT (number of past development activities)	5 or more	5
	4	4
	3	3
	1 - 2	2
	0	1
2. EXISTING DEVELOPMENT (number of present development activities)	5 or more	5
	4	4
	3	3
	1 - 2	2
	0	1
3. POTENTIAL DEVELOPMENT (number of potential development activities)	4 or more	5
	3	4
	2	3
	1	2
	0	1
4. RECREATION ACTIVITIES (number of established or potential recreational activities)	4 or more	5
	3	4
	2	3
	1	2
	0	1

[a] *Criteria developed in same way as those in Table 2.*

Table 3 Indices Related to Human Influence

Existing Development

The existing development index is based upon the number of present land use activities in a particular land unit. For the whole study area, the relative intensity of present development pressures can be estimated from this information, facilitating the selection of possible conservation options on a unit by unit basis.

Potential Development

The potential for future development in a biotic land unit is suggested by the number of kinds of land use activities which are likely to occur in the unit. For each unit this index allows the prospects for conservation to be weighed and related decisions to be guided within the context of possible land uses.

Recreation Activities

The degree of human influence from recreation is measured using criteria which apply only to established and/or potential recreational activities since in most cases insufficient documentation or field evidence exists on past activities. Raw data which discriminate between present and potential recreational activities are given equal weighting during the process, the scoring being based on the assumption that many established recreation areas have a high potential for further recreational development. When viewed in relation to biotic resources, this information on recreation activities could be applied to the identification of possible conflict areas. In these areas, planning decisions related to the type and intensity of recreational activities should be made only after more detailed information on biotic and cultural resources becomes available.

EVALUATION OF BIOTIC LAND UNITS

As explained below, the evaluation of both "Biotic Significance" and "Degree of Human Influence" for each biotic land unit involves the adding and subsequent ranking of score values from the various interpretive indices. Steinitz et al. (1969), after comparing sixteen current resource survey methods, summed up the often-repeated danger of combining, with equal importance, such diverse factors:
" . . . summary evaluations . . . combine what is, in essence, apples and oranges" (p.113). However, they concluded that in spite of the illogical nature of this approach it:

> . . . when properly used, would seem to be a major convenience and indeed, given the manual methods that many of the studies use . . . , the consolidation of many variables into fewer is absolutely necessary. (Steinitz et al., 1969, p.113)

Two additional considerations, the large size of ESAs and, within them, the high number of biotic land units[25] to be systematically evaluated, lend further support to the adoption of the score adding/ranking approach to evaluation.

Biotic Significance

The objective of this part of the evaluation is to distinguish between biotic land units with respect to their relative biotic significance - high, medium or low. For each unit, three types of information are considered: the rank of the total indices scores, the degree to which the unit overlaps with special habitat zones; and biotic features or processes of exceptional interest.

For each unit, evaluation involves a systematic yet flexible process beginning with the summing of score values for community diversity, uniqueness, recoverability, faunal diversity and faunal dependence. Fire susceptibility scores are excluded from this step since generalizations concerning the biological "value" of fire cannot be made[26] (see Postscript, p. 88). Totals are then converted into ranks whose range in turn is divided into

----

[25] One hundred and forty-three biotic land units were delineated for the Bennett Lake/Carcross Dunes/Tagish Lake ESA in the Yukon.

[26] Planning implications related to fire susceptiblity are best determined

three equivalent groups corresponding to high, medium and low biotic significance.

In borderline situations, where a unit is evaluated as one rank below that required for a high or medium rating, its rating may be adjusted in light of the number of special habitat zones with which the unit overlaps. The evaluation rating is raised from low to medium, or medium to high if at least 1/4 of the unit overlaps with two or more special habitat zones.

The final step in the evaluation process involves identifying land units of exceptional biotic significance. These units are assigned a high evaluation rating, regardless of their score totals and rank, and must include such features as ungulate winter range of regional or territorial significance, an assemblage of rare plant or animal species (endemic, relict, endangered, biogeographically peripheral), an unusual abundance of individuals of a particular species, a salt lick, or a site of current biological research or classic past research.

Degree of Human Influence

The objective of this (optional) part of the evaluation process is to distinguish between biotic land units with respect to the degree to which human activities collectively have influenced or may influence biotic resources. In contrast to the biological significance discussed above, this evaluation rating is determined solely through the addition and subsequent ranking of index scores (after Hrabi et al., 1979).

DATA PRESENTATION

Following the preceding discussion of the overall ABC resource survey method and the specific techniques for the biotic component, attention can now be turned to the presentation of the data and results generated by it. Some of the tables and figures presented as examples are specific to the Bennett Lake/Carcross Dunes/Tagish Lake ESA in south-central Yukon Territory, but they illustrate a system for presenting biotic information for use in planning that can be applied to any of northern Canada's ESAs or to parks and related reserves in other areas.

Results can be presented in various formats which allow their consideration at different levels of detail and analysis depending on the objectives and/or technical background of the user. The various data communication formats (Table 4) are meant to display basic characteristics of an ESA's wildlife and vegetation resources. They also provide a means for correlating wildlife species to habitats for ready use in biotic resource

----

on a unit-by-unit basis, depending on such factors as the unit's proximity to urban centres and roads and the potential effects of fire on the unit's wildlife habitat.

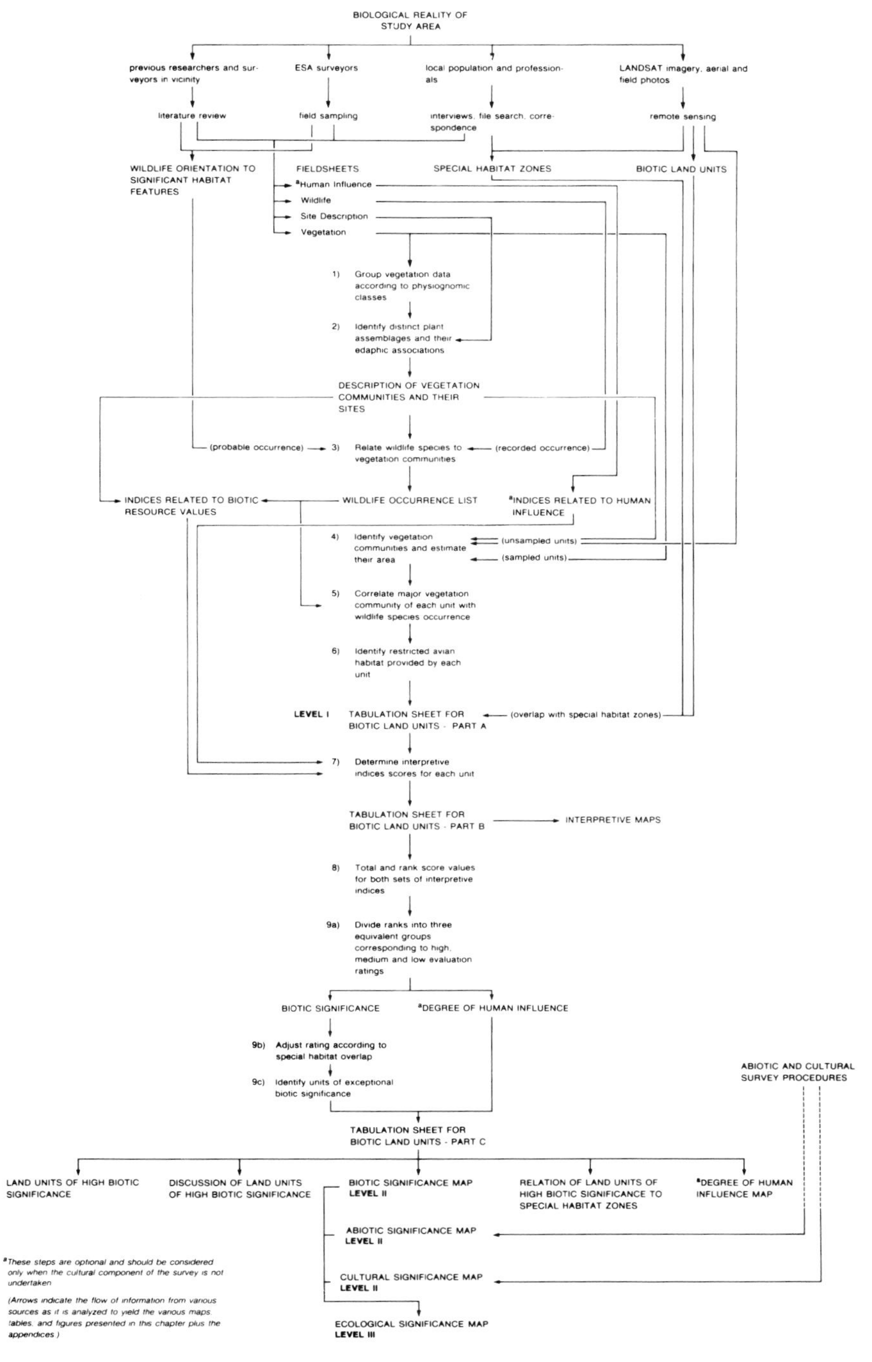

Figure 4 Relationship Between Data Communication Formats

MAP, FIGURE OR TABLE	INFORMATION CONTENT	INFORMATION SOURCE	PURPOSE
FIELDSHEETS (Appendix D)	- descriptions of site (elevation, slope, etc.), vegetation communities (major and minor), wildlife occurrence, significant habitat features, and evidence of human influence	- field sampling, interviews, file search, literature review	- describes outstanding biotic characteristics of major vegetation community - provides summary description of the range of biotic resources within pre-selected biotic land units plus an indication of degree of human influence on these resources
WILDLIFE ORIENTATION TO SIGNIFICANT HABITAT FEATURES (Appendices I & J)	- lists in phylogenetic order the common names of 100 wildlife species occurring in the southern Yukon and relates them to known habitat preferences	- information was obtained from selected references on wildlife in northwestern Canada and from input from local wildlife experts	- aids field recognition of significant habitat features - for determining probable occurence of wildlife species in particular vegetation communities or habitat situations
BIOTIC LAND UNIT MAP (Figure 5)	- displays landmark areas and their constituent biotic land units with corresponding codes	- pre-typing of biotic land units from infra-red LANDSAT image - final boundary delineation from aerial and field photographs	- when used in conjunction with Data Tabulation Sheets (below) provides geographically referenced description of study area's biotic resources
SPECIAL HABITAT ZONE MAP (Figure 6)	- displays riparian zones, cliff/talus/cave complexes, snow-free zones, waterfowl staging areas, fish spawning areas, as well as ungulate winter range, rearing grounds, and migration routes	- combination of remote sensing plus consultation with local residents and wildlife experts	- highlights areas which provide habitat functions of special importance to wildlife particularly large, wide-ranging mammals
DESCRIPTION OF VEGETATION COMMUNITIES AND THEIR SITES (Table 6)	- description and site conditions of vegetation communities within study area (recognized at scale of 1:125,000)	- identification of distinct plant assemblages and their edaphic associations from field sheet information	- provides inference tool for extrapolating vegetation data to non-sampled sites - used as expanded legend for Data Tabulation Sheets - Part A (below) - facilitates recognition of communities by resource managers and land use planners - allows formulation of vegetation component of Indices Related to Biotic Resource Values (below)
WILDLIFE OCCURENCE LIST (Table 7)	- relates wildlife species to vegetation communities in terms of their recorded or probable occurence	- recorded and probable occurence derived from Fieldsheets and Wildlife Orientation to Significant Habitat Features respectively (above)	- provides inference tool for extrapolating wildlife data to non-sampled biotic land units - allows formulation of wildlife component of Indices Related to Biotic Resource Values (below)
INDICES RELATED TO BIOTIC RESOURCE VALUES (Table 2)	- criteria for interpretation of vegetation and wildlife data into ordinal scores reflecting selected biotic resource values: Community Diversity, Uniqueness, Recoverability, Fire Susceptibility, Faunal Diversity, and Faunal Dependence	- established on basis of Description of Vegetation Communities and Their Sites and from Wildlife Occurrence List	- provides detailed interpretation system for the scoring of each biotic land unit with respect to particular biotic resource values
INDICES RELATED TO HUMAN INFLUENCE (Table 3)	- criteria for interpretation of data related to past, present and future human activities into ordinal scores reflecting degree of human influence: Past Impact, Existing Development, Potential Development, Recreation Activities	- established on basis of field sampling results, literature review, and remote sensing	- provides general interpretation system for the scoring of each biotic land unit with respect to particular types of human influence through time
DATA TABULATION SHEET FOR BIOTIC LAND UNITS (Table 5)	Part A - lists for each biotic land unit: its code, total area, major and minor vegetation communities and their area, sparsely and non-vegetated sites and their area, bird species for which restricted habitats are provided, and Special Habitat Zones located within unit; also lists, through time, presence or absence of seven types of development and eight types of recreation activities	- compilation of raw data on biotic land units	- provides convenient storage and reference format for raw data - to be used in conjunction with Biotic Land Unit Map (above) Description of Vegetation Communities and Their Sites (above), and Physiognomic Classes for Vegetation Communities
	Part B - lists for each biotic land unit: the scores for Indices Related to Biotic Resource Values and for Indices Related to Human Influence (above)	- compilation of interpreted data on biotic land units	- provides convenient storage and reference format for interpreted data - used in conjunction with Biotic Land Unit Map (Above)
	Part C - lists for each Biotic Land Unit: the summed total of the scores and its numerical rank, an 'exceptional biotic value' designation if applicable, plus a rating for Biotic Significance, and Degree of Human Influence	- for each type of index scores are summed independently, their totals ranked and then divided into three equal groups corresponding to high, medium and low Biotic Significance ranks and to high, medium or low Degree of Human Influence ranks; Biotic Significance then adjusted in relation to the degree of overlap with Special Habitat Zones and/or to the presence of exceptional biotic features or processes	- provides convenient storage and reference format for evaluation results - used in conjuction with Biotic Significance Map (below)

Table 4 Communication Formats

INTERPRETIVE MAPS (optional)	- separate maps displaying biotic land units in relation to their score values for each of the interpretive indices	- derived from Data Tabulation Sheet - Part B (above)	- provides geographic reference showing relative Uniqueness, Community Diversity, Faunal Diversity, etc. of biotic land units
BIOTIC SIGNIFICANCE (Figure 7)	- displays biotic land units as rating high, medium or low with respect to Biotic Significance	- derived from Data Tabulation Sheet - Part C (above)	- provides geographic reference showing relative biotic significance of biotic land units
MAP OF LAND UNITS OF HIGH BIOTIC SIGNIFI-CANCE (Figure 8)	- displays biotic land units of high Biotic Significance	- derived from Biotic Significance Map (above)	- highlights biotic land units deserving special attention in land use planning due to biotic resource values - used in conjunction with Discussion of Land Units of High Biotic Significance (below)
DISCUSSION OF LAND UNITS OF HIGH BIOTIC SIGNIFICANCE (Table 8)	- lists each biotic land unit with a high Biotic Significance rating in relation to: reasons for high biotic significance, degree of human influence, research needs and opportunities, and immediate management recommendations	- interpretive indices scores used as points of discussion; supplementary information from field notes and personal reflection	- highlights unique biotic values and management considerations of land units of high Biotic Significance - used in conjunction with Map of Land Units of High Biotic Significance (above)
RELATION OF LAND UNITS OF HIGH BIOTIC SIG-NIFICANCE TO SPECIAL HABITAT ZONES (optional)	- displays extent of geographic overlap between land units of high biotic significance and special habitat zones	- overlay of respective maps (above)	- promotes the reconsideration of special habitat zones in light of evaluation results
DEGREE OF HUMAN INFLU-ENCE MAP (optional)	- displays biotic land units as rating high, medium, or low with respect to Degree of Human Influence	- derived from Data Tabulation Sheet - Part C (above)	- provides geographic reference for weighting biotic resource conservation strategies in relation to general human influence

management and land use planning. As is evident from Figure 4, these formats are interrelated and are built in stages with many interlocking parts. The major steps involved in the development and organization of results are also presented in this figure.

In terms of the structure of any proposed report, it is suggested that the first section present a descriptive overview of the study area, highlighting its main features and processes under the headings: *Abiotic Setting, Biotic Setting* and *Cultural Setting*.[27] In the next section, it is suggested that specific Landmark areas should be discussed with respect to their biotic resources. The third section would summarize the evaluation results and their planning implications through a discussion of each land unit of high biotic significance. The most indepth consideration of the results is afforded by the *Data Tabulation Sheet for Biotic Land Units* (Table 5) and its associated tables (e.g. Tables 6a, b and c, and Tables 2, 3 and 7) and figures (e.g. Figures 5, 6, 7 and 8), which together, permit the user to examine all of the raw and interpreted data which culminate in the evaluation of each one of the biotic land units.

----

[27] Abiotic and cultural setting overviews need be included in the final report only when these components are not surveyed concurrently with the biotic surveys.

Finally, a number of figures and tables are presented to illustrate major stages of the overall procedure (see Figure 3) and which would be integrated into the structure of a proposed report at the appropriate places. Figures 5 and 6 are provided as examples of maps produced at Level I, showing respectively structure and functional information for the Bennett Lake/Carcross Dunes/Tagish Lakes ESA. Figure 7 represents a Level II or interpreted map, showing high, medium and low significance ranks for Biotic Land Units. In Figure 8, only the high ranking land units are depicted; thus this map is the cartographic product of greatest significance for planning. To understand the high significance ranking, however, recourse to Level I maps is necessary as well as to the types of tabular information contained in Tables 5 and 8.

QUALIFYING REMARKS

Repeated application of the procedures described above suggests that they provide a relatively rapid and systematic means of acquiring, analyzing and presenting biotic resource data. However, the accuracy of two of the raw data variables recorded in the field should be qualified. First, because of difficulties in detecting clear evidence of past fires in trees other than lodgepole pine *(Pinus contorta* scars easily), acquired data may not accurately reflect the relative fire periodicities among different types of forests (Appendix E, section 8f). Second, observed variations in the character of frost action within alpine/arctic vegetation communities indicate that the causal relationship between the number of types of frost-originated features and the degree of frost action may not be sufficiently strong in some cases to justify inferences concerning recoverabilty of vegetation. Therefore, the feasibility of adopting more accurate indicators of frost intensity should be explored. In the meantime, if this procedure is adopted, results pertaining to recoverability in alpine/arctic areas should be evaluated with this consideration in mind.

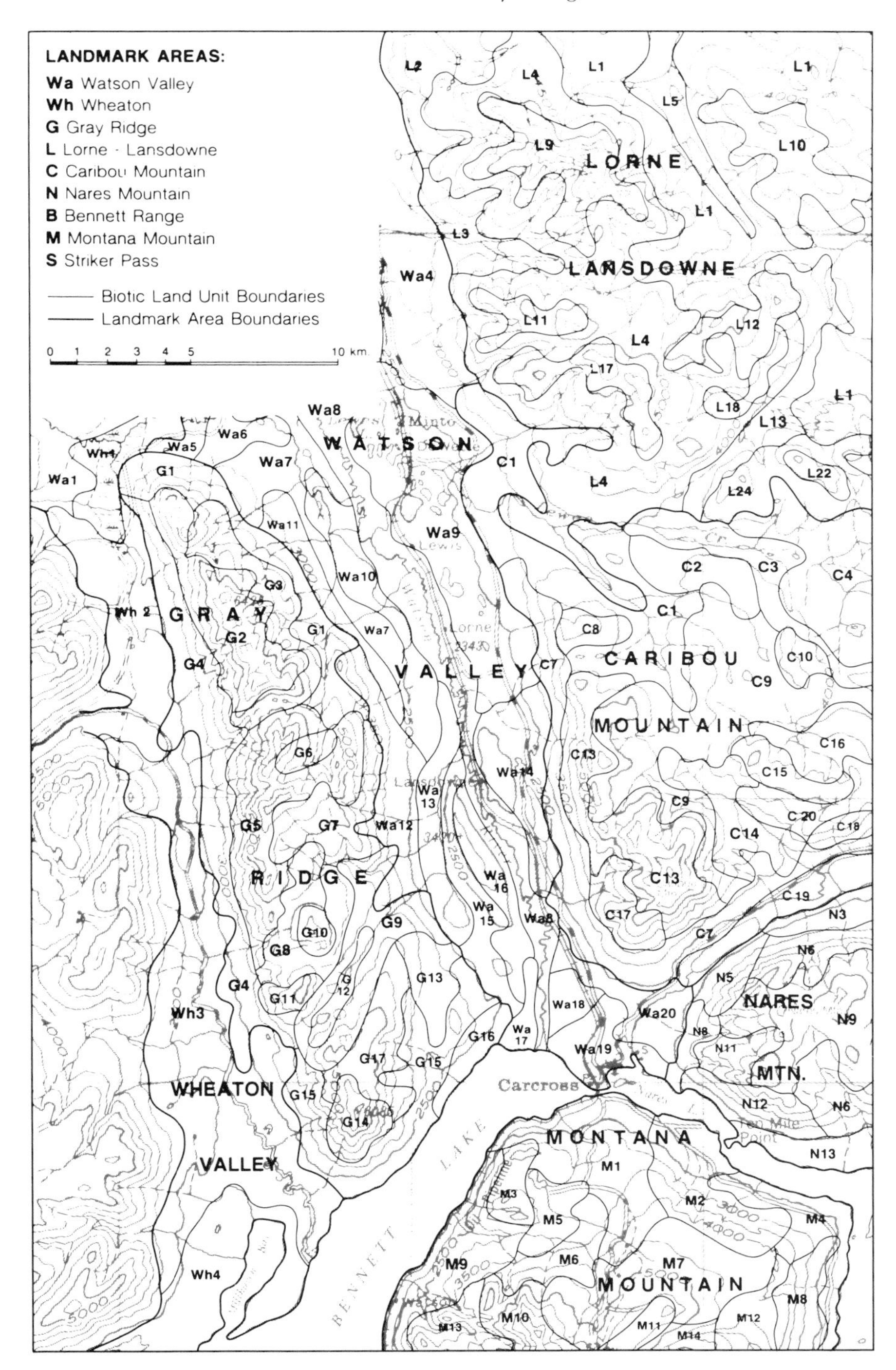

Figure 5 Biotic Land Units

LANDMARK AREA	BIOTIC LAND UNIT (S. SAMPLED)	TOTAL AREA (km²)	A RAW DATA: MAJOR VEGETATION COMMUNITY		MINOR VEGETATION COMMUNITY		SPARSELY OR NON-VEGETATED		RESTRICTED AVIAN HABITAT	SPECIAL HABITAT ZONE	HUMAN-RELATED PROCESSES INFLUENCING VEGETATION AND/OR WILDLIFE WITHIN BLU. (PAST: -, PRESENT: o, FUTURE: +)														COMMENTS	B INTERPRETED DATA: INDICES RELATED TO BIOTIC RESOURCE VALUES							INDICES RELATED TO HUMAN INFLUENCE					C EVALUATION	
																		RECREATION																					
			Community Code(s)	Area (km²)	Community Code(s)	Area (km²)	Community Code(s)	Area (km²)	Bird Species for Which Restricted Habitat Available within BLU	Special Habitat Zones(s) Overlapped by BLU	Scattered Habitation	Urban Development	Cultivation	Grazing	Forestry	Mining	Transportation	Beach Activities	Canoeing	Camping	Cottaging	Boating	Viewing	Park	Biotic or Cultural Features of Unusual Interest	Uniqueness	Recoverability	Fire Susceptibility	Community Diversity	Faunal Diversity (Mammals/Birds)	Faunal Dependence (Mammals/Birds)	Average Score Total (Excl. Fire Sus.)	Existing Development	Potential Development	Human Impact	Recreation Activities	Score Total	BIOTIC SIGNIFICANCE OF BLU	DEGREE OF HUMAN INFLUENCE WITHIN BLU
MONTANA MOUNTAIN	M1 (S)	50	F1.1a	20	F1.1g F1.1e F3.1a	10 10 10			Bald Eagle Osprey	Adjacent to Staging Area		- o +			- o +	- +	- o +				+	+			- Village Site for Carcross Indian Band	5	2	(4)	3	4/3	1/1	15	3	5	4	3	15	Low	High
	M2 (S)	35	F3.1a	18	F1.1c F1.1g F1.1a	5 9 3			Bald Eagle Osprey	Adjacent to Staging Area	-						- o +				o +	+		+	- Proposed Territorial Campground	3	2	(4)	3	4/4	2/1	14	2	2	3	4	11	Low	Medium
	M3 (S)	12	Wdb	3	F2.1a } Sh2.1d } F1.1d F1.1c	2 2 2	SN8.2 SN1 - } SN3 }	1 2		Cliff/Talus/ Cave Complex													o +			5	3	(2)	5	3/4	2/2	18	1	2	1	1	5	Medium	Low
	M4 (S)	23	F1.1e	15	F1.1c F3.1a	4 4			Bald Eagle Osprey		- o				- +		o +	+		+		+				4	3	(4)	2	4/3	1/1	14	2	3	2	4	1	Low	Medium
	M5 (S)	13	Wt3 - Wt1.2	9	SN8.2 Wt1.1	1 3			Herring Gull Mew Gull Bonapartes Gull, Tree Swallow Osprey Bald Eagle	Snow-free Zone						- +	- o		o +	o +					- Active Beaver Lodges - Mine Tailings	5	4	(1)	3	1/3	4/5	19	2	2	2	3	9	High	Medium
	M6	17	F1.2c } Sh2.1 } F1.1g }	17																						3	3	(3)	2	4/3	2/2	14	1	1	1	1	4	Low	Low
	M7 (S)	60	Sh4.2c	45	B1 Sh2.1b M1.b	5 5 3	SN2	2			-					- +	- o +						o		- Extensive Network of Mining Roads, Exploratory Trails	5	5	(2)	5	2/1	2/3	19	2	3	3	2	10	High	Medium
	M8 (S)	63	F2.1 - Sh2.1d	45	F1.1a	3	SN1 SN1 - SN3	10 5	Arctic Tern Bald Eagle Osprey	Cliff/Talus/ Cave Complex	- o					- o +	o +	+			+	+			- Historic Townsite - Venus Mine	4	2	(3)	4	4/3	3/3	17	3	3	3	3	12	Medium	Medium
	M9	64	F1.1e } F1.1a } F3.1a } F1.2c }	64					Bald Eagle Osprey		o				+		- o +	+			+	+			- White Pass Railway along Lakeshore	2	3	(4)	3	4/4.5	2/1	14	2	3	2	3	10	Low	Medium
	M10	29			B1	3	SN1 - } SN2 - } SN3 - } SN4 - }	26		Cliff/Talus/ Cave Complex																4	1	(1)	3	2/1	5/5	15	1	1	1	1	4	Medium	Low
	M11	19	B1	12	M1.a	2	SN2 - SN4	5								- +									- Exploration Mining Trails - Caribou Mine	4	5	(1)	3	3/2	3/4	18	1	2	2	1	6	High	Low

Sample results from Bennett Lake - Carcross Dunes - Tagish Lake ESA, Yukon

Table 5 Data Tabulation Sheet for Biotic Land Units

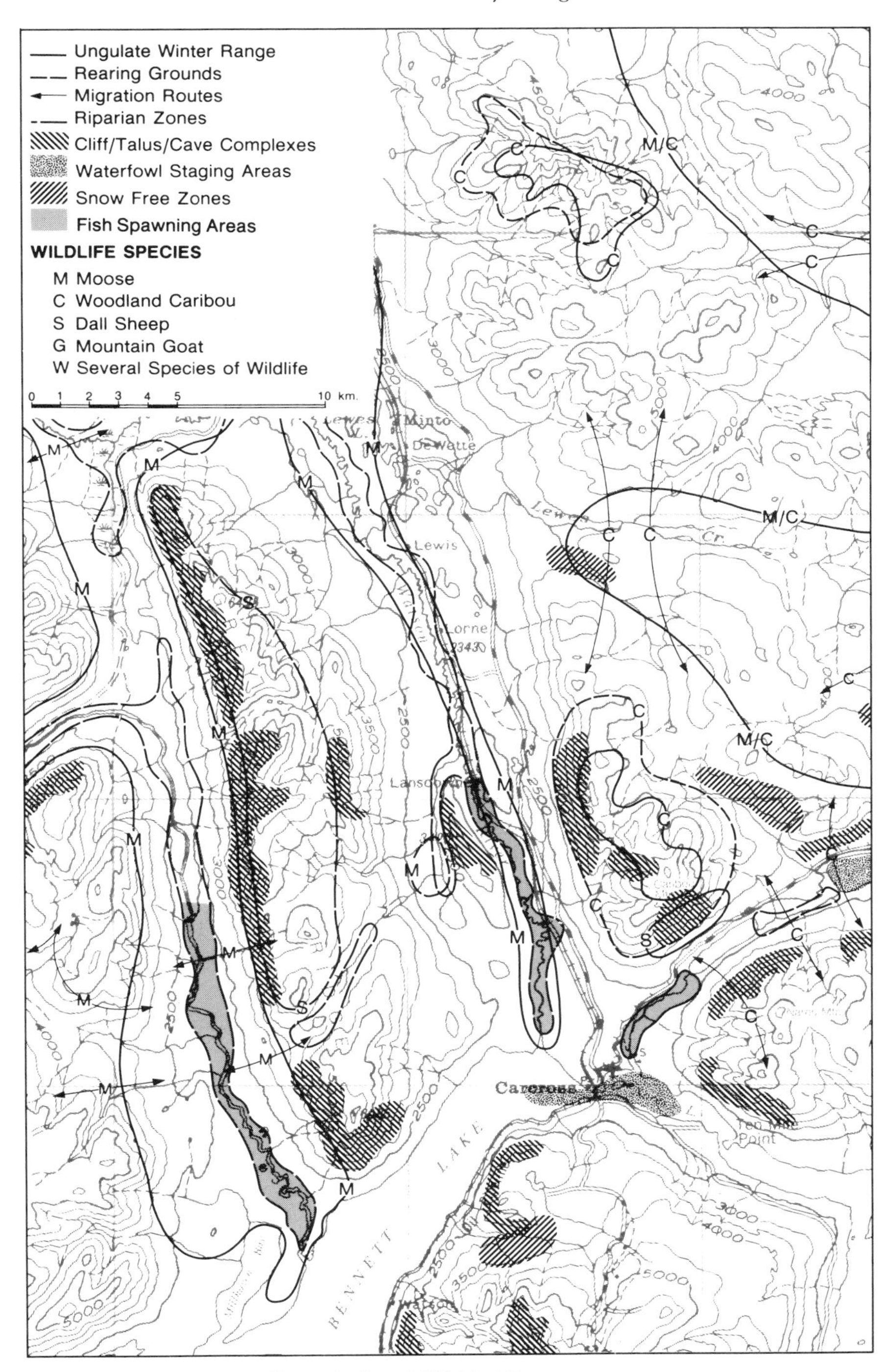

Figure 6 Special Habitat Zones

Table 6a Description of Vegetation Communities and their Sites

	FOREST				BRYOID MAT	WETLAND		SPARSELY OR NON-VEGETATED		
	CLOSED CONIFER		OPEN CONIFER		LICHEN	SHRUB BOG	MARSH			
COMMUNITY NAME (A) = Alpine: upper slope, face and/or apex (SA) = Subalpine: middle slope (L) = Lowland: valley floor and/or lower slope	White Spruce/ Feathermoss/ Forest Lichens (L)	Alpine Fir/Willow Shrub/Feathermoss Forest Lichens (S)	Lodgepole Pine/ Kinnickinnick/ Reindeer Lichens (L)	White Spruce/ Feathermoss/ Reindeer Lichens (L, S)	Alpine and Exposed Alpine Lichens/ Mountain Avens/ Grass/Rock (A)	Bog and Wetland Moss/Willowshrub Birch/Sedge-Rush (L)	Willow/Sedge-Rush/ Water (L)	Unconsolidated Rock Field (A)	Talus (A)	Partially Stabilized Sand Dunes (L)
COMMUNITY CODE **1. COMMUNITY[a] DESCRIPTION**	F1.1a	F1.1b	F1.2a	F1.2b	B1	Wt1.2	Wt3	SN2	SN3	SN6
a) Dominant Species or Species Group (average % cover)	picgla tall (60), picgla low (5), m-fe (60), l-fo (10)	abilas tall (50), abilas low (30), salspp tall (5), m-fe (50), l-fo (20)	pincon tall (45), pincon low (5), arcuva (30), l-rd (20)	picgla tall (40), picgla low (5), 1-rd (20), m-fe (40)	l-al, l-ea } (45), dryint (10), carspp, fesalt } (20)	salspp med (50), betgla med (50), carspp, junspp } (15), m-bg, m-wf } (85)	salspp, tall + med } (75) carspp, junspp } (60)	- vegetation community B1 (20)		- highly variable due to differences in succession, exposure to wind and disturbance
b) Sub-Dominant Species or Species Group(s)	linbor arcrub geoliv pyrasa	pyrsec corcan empnig	corcan epiang graspp	ledpal empnig arcrub	salrot saxopp saxbra m-al	ledpal arcrub equspp	m-wt	l-rk l-ea m-al	l-rk junhor saxtri chrspl	equspp carmar luparc brocum
c) Organic Litter (average % cover of litter group(s))	o-df (10) o-nd, o-cn } (25)	o-df (5) o-nd (5)	o-df (5) o-nd, o-cn } (20)	o-df (5) o-nd, o-cn } (20)	nil	nil	nil	nil	nil	nil
d) Surface Substrate (average % cover)	nil	nil	nil	nil	s-bd, cs (25)	s-wt (5)	s-wt (40)	s-bd, cs (80)	s-cs (90+)	s-sd (85)
e) Canopy Age (years)	150-200	100-150	50-100	50-100	na	na	na	na	na	na
f) Natural Processes Influencing Vegetation and/or Wildlife	Fire: intensive	nil	Fire: intensive, light	Fire: intensive	Frost Action, Nivation	Wildlife: beaver	Wildlife: browsed, beaver	Frost Action, Nivation	nil	Eolian
g) Relative occurrence	17	14	6	1	11	7	3	4	16	2
2. SITE DESCRIPTION										
a) Elevation (m)	660-900	1050-1250	660-940	760-1250	1530-1850	660-1060	650-720	1530-1850	1070-2200	650-730
b) Site Position (macro)	valley floor	middle slope	valley floor to lower slope	lower slope to middle slope	upper slope to apex	valley floor	valley floor	apex	upper slope to apex	valley floor
c) Relief Shape	plain	rolling plain	plain, terraced	fan	hummocky	plain	plain	hummocky	apron	ridged
d) Slope Class	level to gentle	gentle to moderate	level to gentle	moderate	gentle to strong	level	level	moderate	extreme to steep	level to moderate
e) Aspect	none	NW to NE	none	all	all	none	none	all	all	all
f) Exposure Type	nil	nil	nil	nil	wind, frost	frost	nil	wind	wind	wind
g) Ecological Moisture Regime	hygric to mesic	hygric	xeric	mesic	xeric	hygric	hydric	xeric	xeric	xeric
3. RELATION TO OTHER VEGETATION COMMUNITIES										
a) Common Associations	F1.1b F3.1a	F1.1a, F1.2c	nil	nil	SN1, SN2, SN3, SN4	Wt1.1, Wt2, Wt3, F1.1a, SN8.1	Wt1.2, Wt2, SN8.2, SN8.3	SN1, SN3, SN4, B1	SN1, SN2, SN4, Sh4.1	F2.1a, F1.2a
b) Complexes	nil	nil	F1.2a - MIC	nil	nil	Wt1.2 - Wt2, Wt1.2 - Wt3	Wt3 - W1.2	SN1 - SN2 - SN3 - SN4	SN1 - SN3	SN5 - SN6 - SN8.3

[a] *Sample results, as for Table 5.*

SPECIES CODE	SCIENTIFIC NAME	COMMON NAME
abilas	*Abies lasiocarpa*	alpine fir
achbor	*Achillea borealis*	yarrow
anemul	*Anemone multiradiata*	anemone
andsep	*Androsace septentrionalis*	androsace
antros	*Antennaria rosea*	pussytoe
arcuva	*Arctostaphylos uva-ursi*	kinnikinnick
arcrub	*Arctostaphylos rubra*	bearberry
arncor	*Arnica cordifolia*	arnica
artarc	*Artemesia arctica*	sage
artfri	*Artemsia frigida*	sage
betgla	*Betula glandulosap*	shrub birch
bropum	*Bromus pumpellianus*	brome grass
calpur	*Calamagostis purpurascens*	reed bent grass
carmar	*Carex maritima*	sedge
carspp	*Carex spp.*	sedge
castet	*Cassiope tetragona*	arctic white heather
chrwri	*Chrysosplenium wrightii*	chrysosplenium
corcan	*Corus canadensis*	bunchberry
delbra	*Delphineum brachycentrum*	delphineum
dryint	*Dryas integrifolia*	mountain avens
dryspp	*Dryas spp.*	mountain avens
empnig	*Empetrum nigrum*	crowberry
epiang	*Epilobium angustifolium*	fireweed
ericom	*Erigeron compositus*	fleabane
erispp	*Erigeron spp.*	fleabanes
equspp	*Equisetum spp.*	horsetails
fesalt	*Festuca altaica*	bunchgrass
fravir	*Fragaria virginiana*	strawberry
geoliv	*Geocaulon lividum*	bastard toadflax
graspp	*Gramineae spp.*	grasses
juncom	*Juniperus communis*	mountain juniper
junhor	*Juniperus horizontalis*	creeping savin
junspp	*Juniperus spp.*	junipers
ledpal	*Ledum palustre*	Labrador tea
linbor	*Linnaea borealis*	twinflower
luparc	*Lupinus arcticus*	arctic lupine
merpan	*Mertensia paniculata*	bluebell
myoalp	*Myosotis alpestris*	forget-me-not
oxynig	*Oxytropis nigrescens*	oxytropis
papsp	*Papaver sp.*	poppy
pedspp	*Pedicularis spp.*	louseworts
pengor	*Penstemon Gormani*	penstemon
picgla	*Picea glauca*	white spruce

[a]*Sample results, as for Table 5.*

Table 6b Scientific and Common Names Appearing on Table 6a

Table 6c Classes for Vegetation Communities

F FOREST (> 25% tree cover)	Wd WOODLAND (10-25% tree cover)	Sh SHRUBLAND (> 25% shrub cover)	M MEADOW (> 25% grass or forb cover)	B BRYOID MAT (> 25% bryoid cover, treeless and shrubless)	Wt WETLAND (land having water table near, at, or above land surface)	SN SPARSELY OR NON-VEGETATED (0-20% vegetation cover)
t (trees > 5m) l (trees < 5m) 1 Conifer (> 75% of canopy cover) 1 Closed 2 Open 2 Deciduous (> 75% of canopy cover) 1 Closed 2 Open 3 Mixed (conifer and deciduous each contribute 25-75% of canopy cover) 1 Closed 2 Open	1 Conifer (> 75% of canopy cover) 2 Deciduous (> 75% of canopy cover) 3 Mixed (conifer and deciduous each contribute 25-75% of canopy cover)	1 Tall Shrubs (1.5 to 5m) 1 Closed 2 Open 2 Medium Shrubs (0.5 to 1.5m) 1 Closed 2 Open 3 Low Shrubs (0.1 to 0.5m) 1 Closed 2 Open 4 Ground Shrubs (< 0.1m) 1 Closed 2 Open	1 Graminoid (> 50% grass cover) 2 Forb (> 50% forb cover)	1 Lichen Mat (> 50% lichen cover) 2 Moss Mat (> 50% moss cover)	1 Bog 1 Tree Bog (> 10% tree cover) 2 Shrub Bog (> 10% shrub cover) 2 Fen 3 Marsh 4 Swamp 1 Conifer (> 75% of canopy cover) 2 Deciduous (> 75% of canopy cover) 3 Mixed (conifer and deciduous)	1 Consolidated Rock Outcrop (outcrops of massive rock) 2 Unconsolidated Rock Fields (sparse vegetation among boulders or large rocks) 3 Talus 4 Alpine Barrens 5 Active Sand Dunes 6 Partially Stabilized Sand Dunes 7 Eroding Stream Channels 8 Water 1 Shallow Ponds 2 Deep Ponds 3 Lakes 9 Snow and Ice Fields 1 Glacier 2 Rock or Mud Glacier 3 Snow Field 10 Disturbed Land that Would Normally Support Vegetation 1 Clear Cut Forest 2 Land Clearing (borrow pits, etc.) 3 Agriculture 4 Urban/Industrial 11 Naturally Exposed Soil (mineral lake bottoms, marl, etc.)

[a] *Sample classification system for Yukon Territory*

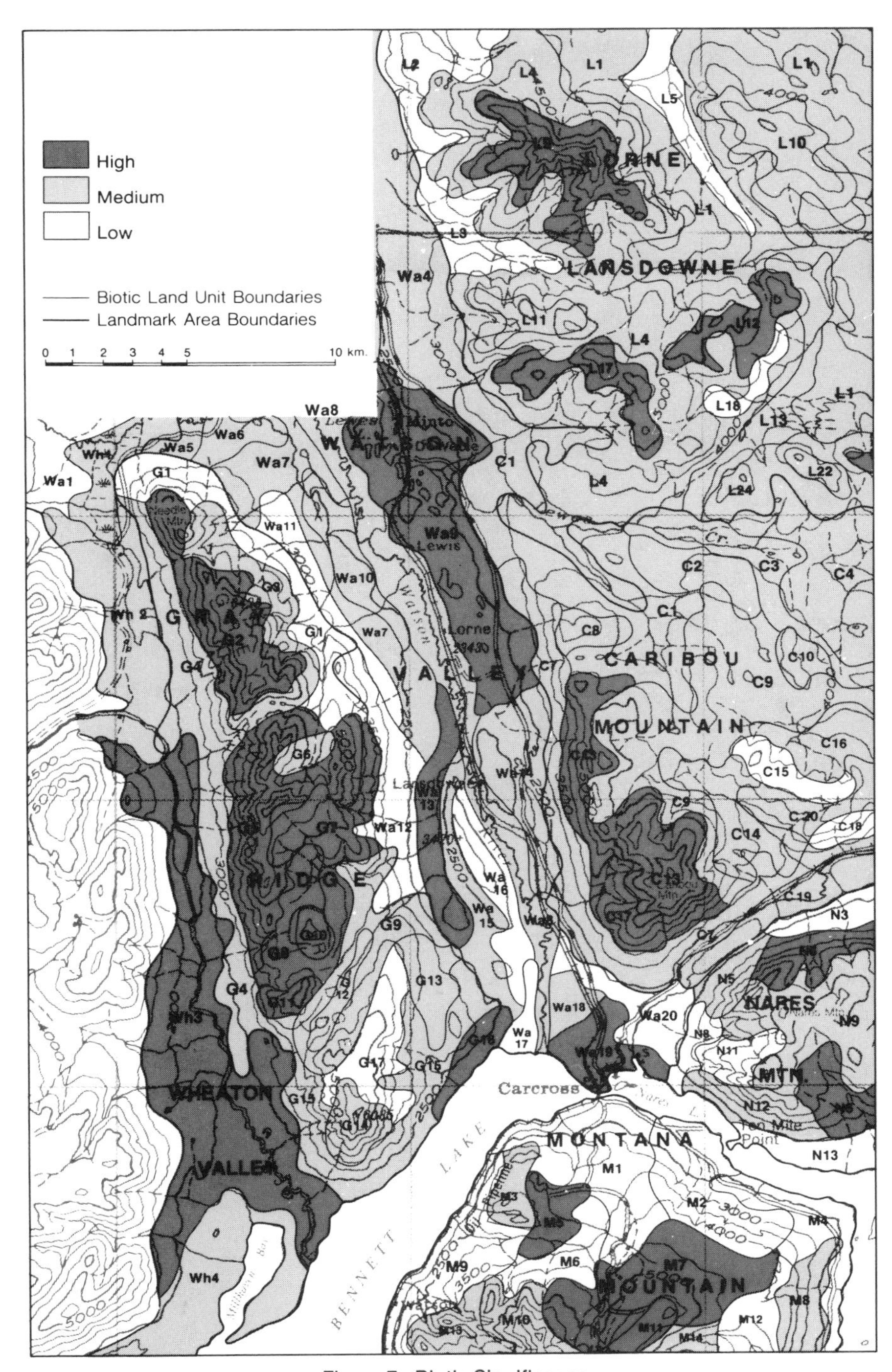

Figure 7 Biotic Significance

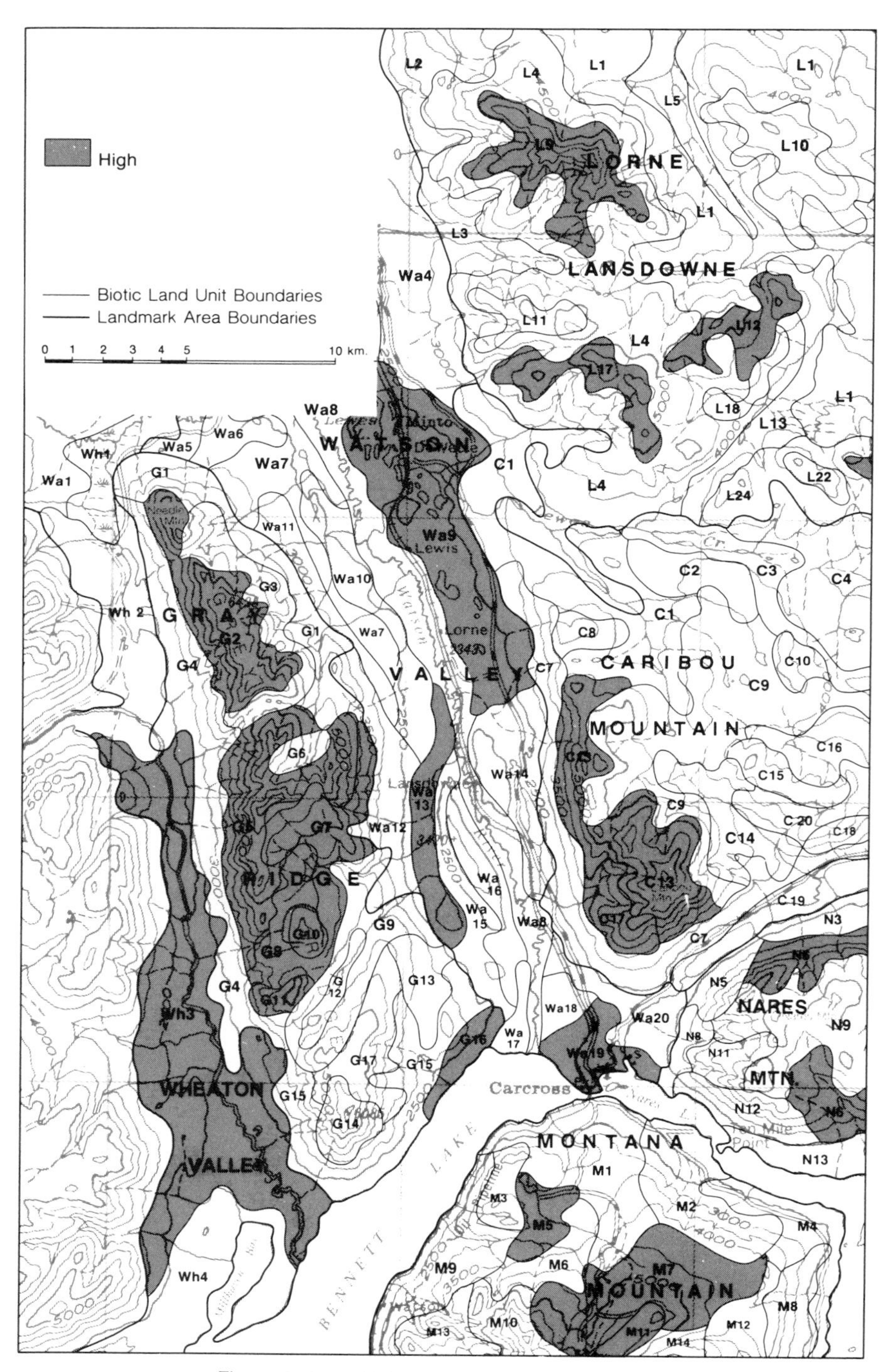

Figure 8 Land Units of High Biotic Significance

		FOREST				BRYOID MAT	WETLAND		SPARSELY OR NON-VEGETATED			
		CLOSED CONIFER		OPEN CONIFER		LICHEN	SHRUB BOG	MARSH				
COMMUNITY NAME (A) = Alpine: upper slope, face and/or apex (SA) = Subalpine: middle slope (L) = Lowland: valley floor and/or lower slope		White Spruce/ Feathermoss/ Forest Lichens (L)	Alpine Fir/Willow Shrub/Feathermoss Forest Lichens (S)	Lodgepole Pine/ Kinnickinnick/ Reindeer Lichens (L)	White Sprice/ Feathermoss/ Reindeer Lichens (L, S)	Alpine and Exposed Alpine Lichens/ Mountain Avens/ Grass/Rock (A)	Bog and Wetland Moss/Willowshrub Birch/Sedge-Rush (L)	Willow/Sedge-Rush/ Water (L)	Unconsolidated Rock Field (A)	Talus (A)	Partially Stabilized Sand Dunes (L)	**VERSATILITY VALUE** (Number of vegetation communities in which species occurs) "V"
COMMUNITY CODE		F1.1a	F1.1b	F1.2a	F1.2b	B1	Wt1.2	Wt3	SN2	SN3	SN6	
MAMMALS												
12. Grizzly Bear						●						10
13. Lynx		●		■	■							9
14. Woodland Caribou		●	●	●	■	●						17
15. Moose		●	●	●	●		●	■				21
16. Dall Sheep									●	●		3
BIRDS												
51. Horned Lark						●					■	5
52. Boreal Chickadee		●	●	●	●							13
53. Red-Breasted Nuthatch		●	■	■	■							13
54. American Robin				■	■		●					11
55. Varied Thrush		●	■									8
No. of Species in Vegetation Community	Mam.	9	7	11	10	7	3	2	5	5	5	
	Birds	17	18	26	26	11	17	18	7	7	5	
Faunal Diversity Score	Mam.	4	3	5	5	3	1	1	2	2	2	
	Birds	3	3	4	4	2	3	3	1	1	1	
% Mammals at or below median "V"		22	14	27	20	43	33	50	80	80	20	
% Birds at or below median "V"		0	11	8	12	45	65	72	57	57	60	
Faunal Dependence Score	Mam.	2	1	2	2	3	3	4	5	5	2	
	Birds	1	1	1	1	4	5	5	5	5	5	

● Recorded occurence: the repeated presence of species in vegetation community as recorded in the field.

■ Probable occurence: the presence of significant habitat features for that species (Appendices I and J) in vegetation community.

[a] *Sample results, as for Table 5.*

Table 7 Wildlife Occurence List

Table 8 Discussion of Land Units of High Biotic Significance

LANDMARK AREA	BIOTIC LAND UNIT	REASONS FOR HIGH BIOTIC SIGNIFICANCE	HUMAN INFLUENCE	IMMEDIATE MANAGEMENT RECOMMENDATIONS	RESEARCH NEEDS AND OPPORTUNITIES
Montana Mountain	M5	- contains largest shrub bog/marsh complex within study area - high diversity of waterfowl and other wetland bird species - large proportion of bird species dependent exclusively on wetlands for nesting habitat - low recoverability of wetland vegetation due to long-accumulated organic soil - suitable habitat for restricted bird species: mew gull, herring gull, Bonaparte's gull, osprey, tree swallow - possible snow-free zone in SW corner of unit - active beaver lodges	- past impact from mine tailings adjacent to lake and tree bog; roads and abandoned structures related to ore crushing site	- rehabilitation of road partially damming north end of lake in order to: a) ensure relative stability of surface water levels and b) to concentrate inevitable recreational access	- monitor effects of tailings run-off on wetland ecosystem and lake - survey unit for possible wolf-denning activity, particularly where kames occur - determine average time of appearance, areal extent, and wildlife use of snow-free zone
	M7	- major community is largest of an infrequently occurring ground shrub community characterized by discrete graminoid and lichen associations - contains several alpine communities including bryoid lichen mat, graminoid/white heather meadows, and sparsely vegetated rock fields - considerable frost action influencing vegetation as indicated by sorted and unsorted circles, hummocks, and solifluction lobes - local sources report traditional use of Montana Mountain area by woodland caribou, present use confirmed by Larsen (1980); this unit may provide best alpine feeding habitat in vicinity	- human impact from mining roads, exploratory trails, and support buildings	- extend closure of caribou hunting in vicinity until further information becomes available on population trends and habitat preferences	- regular fall surveys to monitor distribution and abundance of woodland caribou in Montana area - recoverability studies of road disturbances on alpine vegetation in areas of high frost action - study to investigate vegetation responses to varied microclimates on solifluction lobes (classic example of intimate abiotic-biotic relationship)
	M11	- large proportion of alpine bird and mammal species with low habitat versatility - considerable frost action influencing vegetation as indicated by frequent mud boils, sorted and unsorted circles, solifluction lobes, and stone stripes - uniqueness, mammal diversity and habitat diversity indices all reflect moderate biotic significance	- human impact from mine workings and road exposures causing headward soil erosion and slumping of vegetation on slopes	- environmental impact assessment if renewed mining activity imminent	- studies of rehabilitation or substrate stabilization on or near sites of subsidence

[a] *Sample results, as for Table 5.*

CHAPTER 4

A HINDSIGHT APPRAISAL OF THE ABC SURVEY METHOD

This research was undertaken in response to the need for a method of gathering and synthesizing information pertinent to the future conservation of ecological values in parks, reserves and other environmentally significant areas in Canada's Yukon Territory. To this end, an abiotic, biotic and cultural resource survey approach was developed. Within the context of this approach, specific procedures for the biotic component of the survey were designed, and subsequently have been applied to several ESAs in the Yukon: Frances Lake (Bastedo, 1979); Bennett Lake/Carcross Dunes/Tagish Lake (Bastedo, in press and 1982); Aishihik Lake (Bastedo, in press); Macmillan Pass and Kluane North (Theberge and Fitzsimmons, in preparation). Past experience in applying this methodology was drawn upon to highlight the strengths and weaknesses of the biotic survey procedures in particular and the ABC approach in general.

To aid comparison with other approaches discussed in Chapter 2, to identify opportunities for refinement and to promote elaboration of further methods, this appraisal is framed around the six criteria used earlier, viz. economy, flexibility, replicability, ecological validity, communicability and applicability.

STRENGTHS OF BIOTIC SURVEY PROCEDURES WITH REFERENCE TO THE ABC APPROACH IN GENERAL

"Will the results be used?" is a question that concerns any resource surveyor, regardless of discipline or level of skill. In other words, do the results have sufficient detail, ecological validity, technical soundness and visual impact to be useful in land use planning decisions? Before being tested among management circles, the author can only make predictions on the applicability of results for decision-making.

In this regard, one of the major strengths of the ABC approach is that it yields results in a format favourably suited to the diverse community of potential users. Just as results on abiotic, biotic and cultural resources are, in the proposed approach, interrelated yet separately analyzed and presented, so are there many related yet distinct systems of decision-makers including land managers, administrators, political bodies, public hearing boards, private groups and research scientists. For biotic resources, this means that management opportunities and constraints arising specifically from vegetation and wildlife concerns may be better appreciated and more effectively acted upon by the respective decision-makers.

On the plus side for both the applicability of the biotic results and the economy of methods to acquire them is the fact that maximum advantage is taken of existing data. Thus, results of a season's field work gain additional meaning when synthesized with those of previous studies; as well, the cost-effectiveness of field-sampling is increased by focussing attention on key areas and by avoiding the possiblity of redundant studies. Existing wildlife data are particularly important in this regard. Both abiotic and cultural surveyors take similar advantage of existing data from their respective fields, contributing to the overall applicability and economy of the approach.

During data acquisition, attempts are made to maximize the generalization potential of raw biotic data in order to gain optimum accuracy of results in relation to field-sampling time. For generalizing vegetation characteristics to unsampled units, researchers rely heavily on recorded site descriptions (Tables 6a and 6b) and on field photos (both from the ground and from aircraft); inferences concerning wildlife occurrence are based upon correlations with similar habitat situations found within the study area or referred to in relevant literature (see Appendices I and J). In this way, the more expensive, "incorporative" approach to data acquisition (almost every unit sampled) is avoided.[28]

Replicability of resource survey results is most commonly sacrificed during the interpretation and evaluation stages in which raw data are converted into various indices or measurements meant to reflect and weigh the value of particular resource attributes. Referring to this hazard, Inhaber remarks that "the further we get away from the original [raw] data, . . . the thinner the ice on which we skate" (1976, p.95). Therefore, in the interests of replicability, the possibility of bias entering during these steps is excluded by devising interpretive indices based on quantified and/or non-overlapping criteria and by creating a systematic framework around which to evaluate biotic land units.

Replicability may also suffer during the preliminary delineation and mapping of land units through remote sensing techniques. For this reason, a set of mapping criteria for delineating biotic land units was devised so that inter-observer discrepancies would be minimized. As a test of mapping accuracy, biotic land units for a portion of the Bennett Lake/Carcross Dunes/Tagish Lake ESA were compared to units delineated for the same area from intermediate scale black and white photos using similar mapping criteria (Sauchyn and Bastedo, 1980). Considering the difference in image scale (1:70,000 BxW compared to 1:125,000 LANDSAT) the two sets of units showed a remarkable degree of congruence.

Special habitat zones are mapped using less explicit criteria than biotic land units, yet the degree of overlap among closely related zone types - based on documented as compared to inferred information - suggests that

----

[28] The range of error in extrapolating information could be estimated through ground truthing of several sampled sites.

these zones can be accurately mapped and that they are ecologically valid. For instance, lowland riparian zones, assumed to offer thermal protection to wintering ungulates, often match closely with known moose winter ranges. Similarly, cliff/talus/cave complexes repeatedly correspond with known sheep winter range as would be expected. It should be noted that in all of the cliff/talus/cave complexes sampled, several of the species predicted to use the habitat were discovered. As well, the regular occurrence of staging areas has been corroborated in most cases by local interviews and relevant research. Though the ecological validity of snow-free zones has yet to be tested, the assumptions and recognition criteria for special habitat zones in general appear to be meaningful in this context. Important to note is that the composite picture created by integrating wildlife information from special habitat zones and biotic land units accounts for both sedentary and highly mobile animals.

What do biotic land units mean? As discussed in Chapter 3, biotic land units are not meant to represent discrete ecosystems. Nor do they represent vegetation communities per se. Rather, they represent broad vegetation patterns which, through field investigation, are subclassified into distinct vegetation communities, the largest of which is described in detail. As such, the more appropriate question is, to what degree do biotic land units constitute ecologically valid entities? Rowe (1961) has demonstrated that, as structural assemblages, vegetation communities (or groups of them) in themselves convey little ecological meaning. They become more "ecological" only when related to functions and interactions. As such, biotic land units derive their ecological meaning when vegetation data are correlated with wildlife data so that recurring floral-faunal associations can be discerned. When combined with special habitat zones and restricted avian habitats, particular biotic land units gain further meaning as their specific habitat functions are identified. Following independent analysis, biotic land units gain still further in ecological meaning when viewed in relation to abiotic and cultural land units. These resource categories similarly derive meaning when viewed in relation to biotic land units.

It can be seen that, starting from land units which in themselves may be "non-ecological", the ABC resource survey approach, as designed, culminates in the elucidation of both *intra-* and *inter-*relationships among the three resource categories. For example, the extent to which vegetation patterns are correlated either with man-caused fires or to abiotic factors could be suggested by using this approach. Had resource data been integrated at the data acquisition stage such relationships would remain undetected and potential knowledge valuable for planning and management purposes would be lost.

Only by adopting a process of *inductive* resource analysis can these ecological interactions be revealed. This process stands in contrast to the more popular *deductive* approach in which:

> The environmental variables included in the [survey] coincide with current perceptions of the major elements presumed to act as indicators of ecosystem processes. Within the narrow

> confines of human knowledge concerning the aggregate outcome of environmental interrelationships, the elements chosen are probably indispensable. But potential difficulties [of application] may arise . . . (Cattell, 1977, p.80)

This approach is best exemplified by the notion of an "ecologically significant segment of land" as adopted by proponents of inter-disciplinary resource surveys (see Chapter 3). The author believes that in comparison, an inductive analysis of resources as illustrated by the ABC survey design, offers results which are more ecologically valid, and hence, more applicable to land use planning in environmentally significant areas.

WEAKNESSES OF BIOTIC SURVEY PROCEDURES WITH REFERENCE TO THE ABC APPROACH IN GENERAL

Even without imposing on the land "current perceptions" for recognizing ecosystems, descriptive variables must be chosen over others for reasons of practicality or time constraints, consequently influencing the ecological validity of results. Most notable in this regard is the absence of data on the occurrence of small mammals - shrews, voles, mice, lemmings - and *Mustelidae* - weasels, mink and marten. In most field situations, restrictions in time and personnel disallow the use of traps, necessary for acquiring conclusive data on these animals. Also, due to the relatively narrow habitat requirements for many species (Archibald, 1980), the generalization potential of data at a scale of 1:125,000 is negligible. In contrast, the occurrence of other species, such as wolves, coyotes and foxes, with wide and/or uncertain habitat requirements, may be underestimated in the results due to limitations in field evidence or supportive literature. Finally, the biotic resources of freshwater (and marine) environments remain unaccounted for, again due to constraints in available sampling time. With these gaps in wildlife data, any attempt to describe and compare patterns of energy flow is hampered by missing links in the food chain. Research done by Truett (1979) may provide guidance for the resolution of this problem:

> As the key species and their major food chain organisms were identified, research was structured to study the major physical, chemical, and biological processes that maintained habitats optimal for each key *and* food chain species. A few characteristic processes were found to be important in regulating habitat quality for many of these species. (Truett, 1979, p.358)

With respect to wildlife, future ESA surveyors may choose to concentrate sampling efforts on only those features or processes which are, through their effects on the food chain, important to the support of key species such as summit predators or major prey groups. Results from such studies would be particularly useful in developing management strategies aimed at maintaining food chain stability.

Detracting from the replicability of results is the approach's emphasis on the identification of key areas, namely units of high abiotic, biotic or cultural significance. These units are featured in the results as they are considered to be of prime concern to land use planning yet, in some cases, their identification may depend upon serendipitous findings. Examples from biotic surveys include salt licks and rearing grounds. The location of these biotically significant areas is relatively difficult to predict as they are often discovered by chance or through local knowledge. Discrepancies in information availability (and good fortune) will vary inevitably from place to place or from time to time; therefore, the replicability of results may be questioned. Decision makers may hesitate to apply the results fearing the consequences of action in areas where such information may have been overlooked. To this concern one could reply that, because of the climate of impending development found within and around the majority of ESAs, land use decisions will be made regardless of the amount of information available. The author agrees with Giles (cited by Thomas, 1979, p.12) who writes: "Certainly, [more] research is needed, but while waiting, we need to work with what we have".

Significance is a term which may impede the communicability of survey results since it begs the question, "Significant to what?". Unrelated as it is to any particular land use and based upon variables which have little to do with economic interests, the term may initially seem ambiguous, if not useless, to potential users of the information. Etymologically, significance is defined as: "having or conveying a special meaning; suggestive" (Webster, 1969, p.1688). "Significance", then, in the context of ESAs, should be understood as a pre-defined measure suggesting the relative ecological value of land units; this value is expressed in relation to combinations of natural or human features and processes judged to be important as regulators or indicators of ecological integrity (the maintenance of essential ecological processes and life support systems). In this light, the response to the question, "significant to what?" is: "to planning or management decisions involving the allocation of land uses which may increase, maintain or decrease ecological values as defined above", or in short, if it is assumed that all land uses in some way influence these values: "to *any* land use decisions for which the scale of information is appropriate".

CONCLUDING REMARKS

The conservation of ecological values is just one among many possible land use objectives guiding future land use decisions for ESAs. As in all decision-making processes where several values are involved, conflict is inevitable; in some land units ecological values will be traded off for political, social or economic values derived from incompatible land uses. In these cases, results of the ABC survey approach here presented could be used in deciding where trade-offs should be made and of what they should consist. However, consistent use of survey information in this disjointed fashion could lead to the preservation of only the "most valuable" (highly significant) units, with other units left to destruction (Van de Ploeg and Vlijm, 1978). Projection of this trend into the long term creates the picture of a land fractionated incrementally by a series of land use decisions unconnected in time or intent.

Such a planning scenario represents a misapplication of survey results. Units of high significance - whether abiotic, biotic or cultural - should be considered not as isolated resource pools but rather as key threads in the fabric of ecosystems. In this sense, these units could be agglomerated with other units to which they are linked through interdependent functions or complementary features. The relative significance among neighbouring or overlapping land units would aid in the recognition of such linkages and suggest boundaries for appropriately-sized planning or management areas. In turn, these areas could be allocated to suitable policy classes which include special restrictions on land use within highly significant units. The ability of particular units or groups of units to sustain certain land uses could thus provide the foundation for zoning or other land use control systems for ESAs.

Where conservation of ecological values takes precedence, land use planning decisions should depend upon knowledge, gained from resource survey results, of the land's predicted response to particular types of land use. Since predictive power of results is a function of the degree to which ecological relationships are identified, the proposed approach would gain maximum usefulness for planning following the integration of abiotic, biotic and cultural information.

As stated in Chapter 1, *planning links knowledge to action* - a simple definition for a complex task. For ESAs, details on zoning, boundary delineation and other potential applications of survey results await the fruits of ongoing research. At the time of writing (May, 1982), compatible procedures for the abiotic and cultural components of the survey are being developed; zoning mechanisms are being explored as are other institutional arrangements related to the acquisition and management of ESA lands. In effect, a comprehensive land use planning process for ESAs is being developed. More than an academic exercise, this work is continually inspired by the widespread need to affirm the legitimacy of ecological values in establishing land use priorities.

POSTSCRIPT

Since the completion of this study in May 1982, continuing ESA research at the University of Waterloo has yielded two major modifications to the ABC survey methodology. As can be seen on Figure 9, these modifications constitute additions to the initial model (Figure 3), so they do not in any way alter the underlying concepts and field procedures described in this report.

The first modification involves the crucial stage of interpretation in which raw data are translated into a form which reflects selected resource attributes. We have discovered that interpretive indices fall naturally into

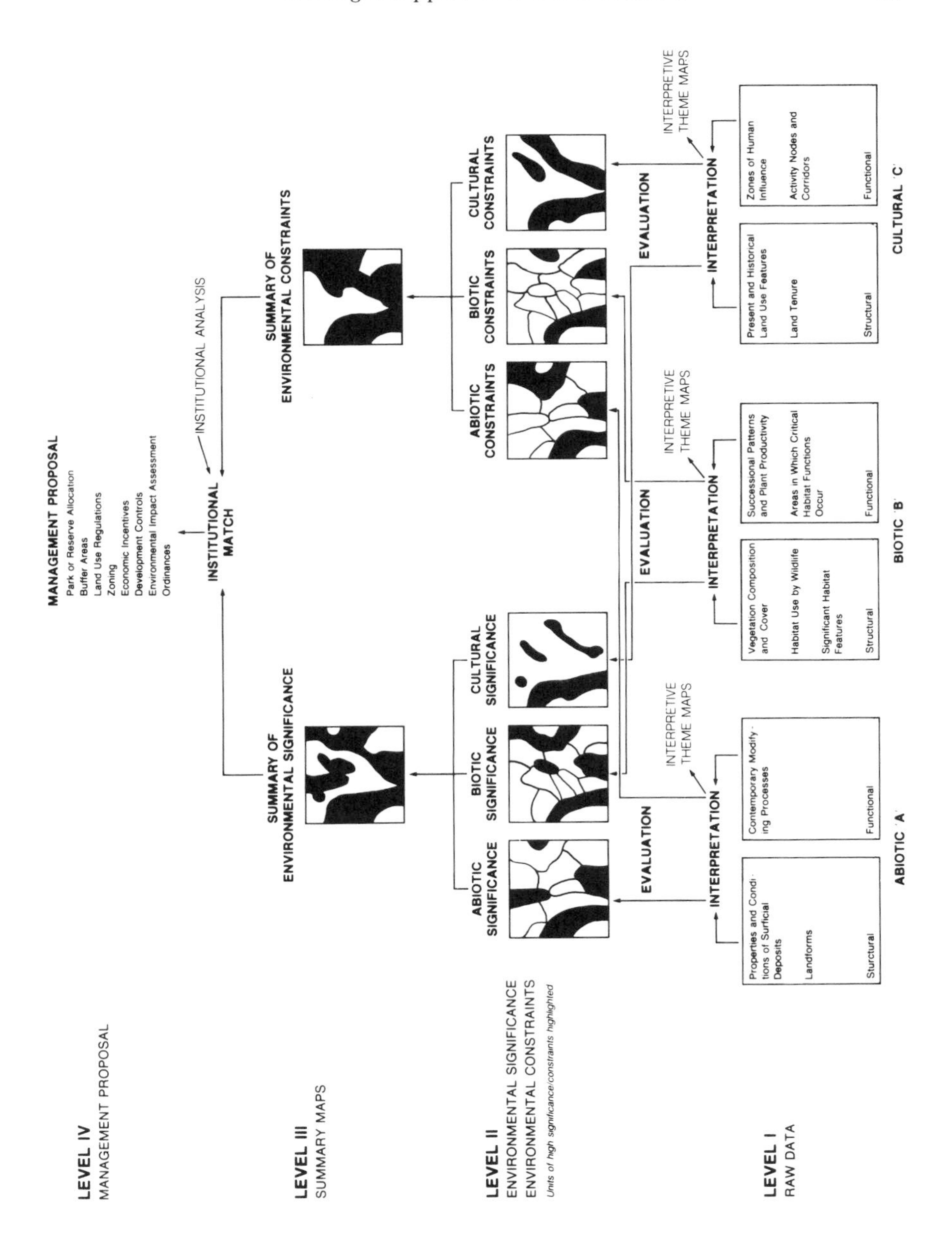

Figure 9 Revised Model of ABC Resource Survey and Planning Approach for Environmentally Significant Areas

two mutually exclusive categories, which we refer to as *environmental significance* and *environmental constraints,* a distinction which helps to minimize problems arising from "adding apples and oranges" (for instance, how to handle information on Fire Susceptibility). The former category includes indices related to wildlife, historic and other resource values while those in the latter category reflect biophysical hazards and sensitivities as well as land use conflicts. Within the biotic component, this means that the indices of Faunal Diversity, Community Diversity and Uniqueness are grouped together to yield the Level II Biotic Significance map, while the indices of Faunal Dependence, Recoverability and Fire Susceptibility are grouped together to yield the Biotic Constraints map. According to the revised model, these maps are integrated with their respective abiotic and cultural maps to produce two summary maps which, together, focus the users' attention upon particular areas where ecological values are concentrated and to view these areas in relation to environmental constraints, whether abiotic, biotic or cultural.

The second modification to the model is the addition of a fourth level of integration, referred to as a *Management Proposal.* The Level IV product shows proposed park or reserve allocations, buffer areas or other land use controls and is arrived at by matching an ESA's distinctive biophysical and cultural attributes with suitable land use and institutional arrangements. Several alternative maps can be drawn up for consideration by the major user groups. Viewed in the context of lower level maps, the Level IV maps could provide the prime focus of discussion within various decision-making forums.

In closing, it is important to note that while we have developed a comprehensive methodology for resource survey and planning which involves steps ranging from the acquisition of abiotic, biotic and cultural data to the selection of appropriate institutional arrangements, the system may be applied in whole or in part (for instance, dealing only with the biotic component to Level II) depending on the objectives and orientation of possible users.

SELECTED GLOSSARY

The following terms deserve full explanation in glossary form as they occur frequently throughout the text and their meanings are crucial to a full understanding of this study.

BIOTIC LAND UNIT

Unit of land surface characterized by a distinct vegetation pattern which corresponds to a particular *LANDSAT* theme (spectral signature). Through field investigation, this pattern is subclassified into distinct vegetation communities, the largest of which is described in detail. Collectively, biotic land units provide a geographic framework with which to describe existing vegetation communities within a study area, as well as to identify faunal assemblages and habitat features which characterize these communities.

CLASSIFICATION

A resource survey step in which the environment is categorized into mapped land units according to a pre-determined set of criteria. This step is usually preceded by *Data Variable Selection* and followed by *Data Acquisition.*

CONSERVATION

A general planning/management term referring to the maintenance of resource use options. As a process, conservation can be broken down into four strategies which correspond to varying degrees of environmental manipulation. From least to greatest degree of manipulation, these conservation strategies are *preservation, protection, managed use* and *restoration* (Naysmith, 1971).

ECOLOGICAL INTEGRATION

Part of the classification process for some resource surveys in which raw or interpreted data are combined to identify "ecological units" through the recognition of interrelationships among resource categories.

EVALUATION

A resource survey step in which interpreted data are combined in such a way as to determine: (a) the relative appropriateness of land units for various uses or (b) the relative worth of land units in terms of some pre-defined value measure.

INTERPRETATION

A resource survey step in which raw data are translated into a form which reflects selected resource attributes and which allows comparison of these attributes among land units. This step is usually preceded by *Data Acquisition* and followed by *Evaluation.*

LANDMARK AREA

Geographic reference area consisting of several biotic land units. These areas are bounded by major drainage systems, mountain ridges or roads and are named after prominent landscape features within their boundaries.

LAND USE PLANNING

The linking of information about an area of land to the action of allocating land units for various kinds and intensities of resource use. As an idealized process, land use planning consists of ten major steps:

1. *Planning Objectives*
2. *Information Requirements*
3. *Data Variable Selection*
4. *Classification*
5. *Data Acquisition*
6. *Data Interpretation*
7. *Evaluation*
8. *Prescription*
9. *Communication*
10. *Zoning or other land use controls.*

LOGISTICAL INTEGRATION

During resource surveys, the integration of research efforts (field schedules, transportation, camps, etc.) and write-up activities among surveyors from several resource disciplines.

PRESERVATION

Conservation strategy involving minimal levels of environmental manipulation so as to safeguard existing features or processes. This strategy may include a spectrum of manipulation, e.g. habitat alteration to preserve successional stage of vegetation (*era* approach) or 'hands-off' management to allow successional changes in vegetation (*evolutionary* approach).

PROCEDURES

Set of specific resource survey techniques presented in sufficient detail as to allow immediate field or laboratory application.

PROTECTION

Conservation strategy involving moderate levels of environmental manipulation. Includes precautionary measures (e.g. land use regulations) which protect features or processes from potentially damaging effects of a specific project or development program.

RESOURCE SURVEY

Study which measures the position and extent of one or more resources within a given area and are presented in cartographic form usually accompanied by a text report.

RESOURCE SURVEY APPROACH

General methodological framework for a particular type of resource survey, the techniques for which are not given in workable detail.

SIGNIFICANCE

Value measure used to determine the degree to which land units deserve priority in land use planning considerations. The term may be applied specifically to abiotic, biotic or cultural resources or to their combined ecological value.

REFERENCES

Alvis, R.J. (1978) "Time-Space and the Inventory of Ecosystems", in *Integrated Inventories of Renewable Natural Resources,* Proceedings of the Workshop, January 8-12, 1978, Tucson, Arizona, pp.299-303, Washington: U.S. Department of Agriculture, Forest Service.

Anonymous (1978) "Parks and Scientific Preserves, Working Group Report", in *Northern Transition,* Vol. 2, Second National Workshop on People, Resources and the Environment North of 60 Degrees, Keith, R.F. and Wright, J.B. (eds.), pp. 228-230, Ottawa: Canadian Arctic Resources Committee.

Archibald, R. (1980) Yukon Wildlife Branch, Whitehorse (personal communication, August 6).

Bailey, R.G., Pfister, R.D. and Henderson, J.A. (1978) "Nature of Land and Resource Classification - A Review", *Journal of Forestry,* 76:650-655.

Banfield, A.W.F. (1974) *The Mammals of Canada,* Toronto: University of Toronto Press.

Bastedo, J.D. (1979) *An Ecological Survey Method for Environmentally Significant Areas in the Yukon with Results from Frances Lake,* Waterloo: University of Waterloo, President's Committee on Northern Studies, Working Paper No. 5.

Bastedo, J.D. (1982) *A Resource Survey Method for Environmentally Significant Areas in the Yukon with Results from the Bennett Lake/Carcross Dunes/Tagish Lake ESA,* Waterloo: University of Waterloo, unpublished M.A. thesis.

Bastedo, J.D. (in press) *Resource Survey of the Bennett Lake/Carcross Dunes/Tagish Lake Environmentally Significant Area, Yukon: An Evaluation of Biotic Resources,* Waterloo: University of Waterloo, President's Committee on Northern Studies.

Bossort, A.K. (1978) Masters thesis proposal, submitted to the Department of Plant Ecology and Institute for Northern Studies, University of Saskatchewan, unpublished typescript.

Canada (1968) *Land Capability for Wildlife - Ungulates,* Edmonton Map Sheet Area, 83 H, Ottawa: Department of Agriculture, Soil Research Institute.

Canada (1972-1981) Department of Indian and Northern Affairs and Department of Fisheries and the Environment, *Land Use Information Series,* Ottawa: Department of Fisheries and Environment, Lands Directorate.

Canadian Wildlife Service (1972) *Snowshoe Hare,* Hinterland Who's Who Series, Ottawa: Environment Canada.

Canadian Wildlife Service (1973a) *Wolf,* Hinterland Who's Who Series, Ottawa: Environment Canada.

Canadian Wildlife Service (1973b) *Caribou,* Hinterland Who's Who Series, Ottawa: Environment Canada.

Canadian Wildlife Service (1973c) *Mountain Sheep,* Hinterland Who's Who Series, Ottawa: Environment Canada.

Canadian Wildlife Service (1977b) *Porcupine,* Hinterland Who's Who Series, Ottawa: Environment Canada.

Canadian Wildlife Service (1977c) *Coyote,* Hinterland Who's Who Series, Ottawa: Environment Canada.

Canadian Wildlife Service (1977d) *Canada Lynx,* Hinterland Who's Who Series, Ottawa: Environment Canada.

Canadian Wildlife Service (1977e) *Moose,* Hinterland Who's Who Series, Ottawa: Environment Canada.

Canadian Wildlife Service (1978a) *Beaver,* Hinterland Who's Who Series, Ottawa: Environment Canada.

Canadian Wildlife Service (1978b) *Red Fox,* Hinterland Who's Who Series, Ottawa: Environment Canada.

Canadian Wildlife Service (1980) *Population Ecology Studies of the Polar Bear in Northern Labrador,* Ottawa: Information Canada.

Cattell, K.M. (1977) *An Evaluation of the Canadian National Parks Zoning System,* Report prepared for Parks Canada.

Chambers, B. (1979) "The Approach to Planning in Yukon", in *Resource Inventory Workshop,* Proceedings, Whitehorse, October 16-18, 1979, pp.73-80, Whitehorse: Government of the Yukon Territory.

Christian, C.S. (1959) "The Concept of Land Units and Land Systems", *Proceedings, Ninth Pacific Science Congr.,* 20:74-81.

Coleman, D.J. (1977) "Environmental Impact Assessment Methodologies: A Critical Review", in *Environmental Impact Assessment in Canada: Processes and Approaches,* Plewes, M. and Whitney, J.B. (eds.), pp.35-39, Toronto: University of Toronto, Institute for Environmental Studies.

Coulombe, H.N. (1978) "Toward an Integrated Ecological Assessment of Wildlife Habitat", in *Integrated Inventories of Renewable Natural Resources,* Proceedings of the Workshop, January 8-12, 1978, Tucson, Arizona, pp.5-23, Washington: U.S. Department of Agriculture, Forest Service.

Cowan, McTaggart I. (1977) *Natural Resources Research in Canada's National Parks: An Evaluation,* Ottawa: Parks Canada, Natural Resources Branch.

Crampton, C.B. (1973) *Landscape Survey in the Upper and Central MacKenzie Valley,* Task Force on Northern Oil Development Report, No. 73-8, Ottawa: Information Canada.

Crowe, K.J. (1974) *A History of the Original Peoples of Northern Canada,* Montreal: McGill-Queen's University Press.

Day, J.H. (1962) *Reconnaissance Soil Survey of the Tahkini and Dezadeash Valleys in the Yukon Territory,* Ottawa: Canada Department of Agriculture, Research Branch.

Day, J.H. (1972) *Soils of the Slave River Lowland in the Northwest Territories,* Sheet 1, Ottawa: Department of Energy, Mines and Resources, Surveys and Mapping Branch.

Day, D.L. (1978) "Organization, Storage and Retrieval of Bio-physical Information on a Site, Regional and National Basis", Discussion paper prepared for 1978 Superintendent's Conference, Parks Canada.

Department of Indian Affairs and Northern Development (1972) *Research Under the Environmental-Social Program Northern Pipelines,* Task Force on Northern Oil Development, Report No. 72-1, Ottawa: Department of Indian Affairs and Northern Development.

Department of Indian and Northern Affairs (1977) *Biophysical Inventories: Concepts, Methodology, Applications, Parks Canada, Atlantic Region,* Ottawa: Department of Indian and Northern Affairs.

DeVos, A. and Mosby, H.S. (1971) "Habitat Analysis and Evaluation", in *Wildlife Management Techniques,* 3rd edition, Giles, R.H. (ed.), pp.135-172, Washington: The Wildlife Society.

Dickinson, D.M. (1978) "Northern Resources: A Study of Constraints, Conflicts, and Alternatives", in *Northern Transitions,* Vol. 1, Northern Resource and Land Use Policy Study, Peterson E.B. and Wright, J.B. (eds.), pp.253-316, Ottawa: Canadian Arctice Resources Committee.

Dorney, R.S. (1976) "Biophysical and Cultural-Historic Land Classification and Mapping for Canadian Urban and Urbanizing Landscapes", in *Ecological (Biophysical) Land Classification in Urban Areas,* Ecological Land Classification Series, No. 3, Wiken, E.B. and Ironside, G.T. (eds.), pp.55-61, Ottawa: Environment Canada, Lands Directorate.

Dorney, R.S. (1978) "Philosophical and Technical Principles for Identifying Environmental Planning and Management Strategies", in *Northern Transitions,* Vol. 2, Second National Workshop on People, Resources and the Environment North of 60 Degrees, Keith, R.F. and Wright, J.B. (eds.), pp.144-149, Ottawa: Canadian Arctic Resources Committee.

Dorney, R.S. and Hoffman, D.W. (1979) "Development of Landscape Planning Concepts and Management Strategies for an Urbanizing Agricultural Region", *Landscape Planning,* 6:151-177.

Douglas, G.W. (1974) "Montane Zone Vegetation of the Alsek River Region, Southwest Yukon", *Canadian Journal of Botany,* 52(12):2505-2532.

East, K.M., Day, D., LeSauteur, D., Stephenson, W.M. and Charron, L. (1979) "Parks Canada Application of Biophysical Land Classification for Resources Management", in *Applications of Ecological (Biophysical) Land Classification in Canada,* Ecological Land Classification Series, No. 7, Rubec, C.D.A. (ed)., pp.209-220, Ottawa: Environment Canada, Lands Directorate.

Environment Canada (1978) *The Canada Land Inventory: Objectives, Scope and Organization,* Ottawa: Environment Canada, Lands Directorate.

Environment Canada, (1980) *Ecological Land Survey Guidelines for Environmental Impact Analysis* (Preliminary Draft), Ottawa: Environment Canada, Lands Directorate, Environment Management Service.

Erskine, A.J. (1977) *Birds in Boreal Canada: Communities, Densities and Adaptations,* Canadian Wildlife Service Report Series, No. 41, Ottawa: Fisheries and Environment Canada.

Feachem, R.G. (1977) "The Human Ecologist as Superman?" in *Subsistence and Survival, Rural Ecology in the Pacific,* Bayliss-Smith, T.P. and Feachem, R.G. (eds.), pp.3-10, London: Academic Press.

Frayer, W.E., Davis, L.S. and Risser, P.G. (1978) "Uses of Land Classification", *Journal of Forestry,* 76(10):647-649.

Geist, V. (1974) *Mountain Sheep: A Study in Behaviour and Evolution,* Chicago: University of Chicago Press.

Gehlbach, F.R. (1975) "Investigation, Evaluation and Priority Ranking of Natural Areas", *Biological Conservation,* 8:79-88.

Gimbarzevsky, P. (1978) "Land Classification as a Base for Integrated Renewable Resources Inventories", in *Integrated Inventories of Renewable Natural Resources,* Proceedings of the Workshop, January 8-12, 1978, Tucson, Arizona, pp.169-177, Washington: U.S. Department of Agriculture, Forest Service.

Godfrey, W.E. (1976) *The Birds of Canada,* Ottawa: Supply and Services Canada.

Goldsmith, F.B. (1975) "The Evaluation of Ecological Resources in the Countryside for Conservation Purposes", *Biological Conservation,* 8:89-96.

Grigoriew, P. (1980) *Natural Area Preservation: An Inquiry and a Comparative Case Study,* Toronto: York University, unpublished M.E.S. thesis.

Grigel, J.E., Lieskovsky, R.J. and Kiil, A.D. (1971) *Fire Hazard Classification for Waterton Lakes National Park,* Edmonton: Department of the Environment, Canadian Forest Service.

Hans, B.L. (1981) "Proposed Analytical Framework for Cultural Resource Surveys of the Yukon Environmentally Significant Areas", Waterloo: University of Waterloo, unpublished typescript.

Hettinger, L., Janz, A. and Wein, R.W. (1973) *Vegetation of the Northern Yukon Territory,* Biological Report Series, Vol. 1, Calgary: Arctic Gas.

Hills, G.A. (1961) *The Ecological Basis for Land Use Planning,* Toronto: Ontario Department of Lands and Forests, Research Report No. 46.

Hirsch, A., Krohn, W.B., Schweitzer, D.L. and Thomas, C.H. (1979) "Trends and Needs in Federal Inventories of Wildlife Habitat", in *Transactions of the Forty-Fourth North American Wildlife and Natural Resources Conference,* Toronto, March 24-28, 1979, pp.340-368, Washington: U.S. Fish and Wildlife Service.

Hoefs, M. (1976) "Birds of the Kluane Game Sanctuary, Yukon Territory and Adjacent Areas", *Canadian Field Naturalist,* 87:345-355.

Hoefs, M., Lortie, G. and Russel, D. (1977) "Distribution, Abundance and Management of Mountain Goats in the Yukon", in Samuel, W. and MacGregor, W.G. (eds.), *Proceedings of the First International Mountain Goat Symposium,* Victoria: Fish and Wildlife Branch of the British Columbia Government.

Holland, S.S. (1976) *Landforms of British Columbia: A Physiographic Outline,* Victoria: British Columbia Ministry of Energy, Mines and Petroleum Resouces, Bulletin 48.

Holroyd, G.L. (1980a) "Inventorying Wildlife in Banff and Jasper National Parks and Possible Applications in the Yukon", in *Resource Inventory Workshop,* Proceedings, Whitehorse, October 16-18, 1979, pp.187-193, Whitehorse, Government of the Yukon.

Holroyd, G.L. (1980b) "The Biophysical Wildlife Inventory of Banff and Jasper National Parks", in *Land/Wildlife Integration,* Ecological Land Classification Series, No. 11, Taylor, D.G. (ed.), pp.37-43, Ottawa: Environment Canada, Lands Directorate.

Holroyd, G.L., Van Tighem, K.J., Skeel, M.A. and Kansas, J.L. (1979) *The Biophysical Inventory of Jasper and Banff National Parks,* Interim Report, 1978-1979, Banff, Jasper: Canadian Wildlife Service.

Hrabi, C.A., Diebolt, R. and Kyle, H. (1979) "Priority Ranking of Environmentally Sensitive Areas", Waterloo: University of Waterloo.

Inhaber, H. (1976) *Environmental Indices,* New York: Wiley-Interscience.

Jaques, D.R. (1976) "Winter Alpine-Sub Alpine Wildlife Habitat in the Southern Rocky Mountains of Alberta", Calgary: University of Calgary, Environmental Sciences Centre, unpublished typescript.

Jurdant, M., Belair, J.L., Gerardin, V. and Wells, R. (1974) "Ecological Land Survey", in *Canada's Northlands,* Ecological Land Classification Series, No. 0, Romaine, M.J. and Ironside, G.R. (eds.), pp.23-30, Ottawa: Environment Canada, Lands Directorate.

Kananaskis Centre (1979) *Proceedings of the Kananaskis Seminar on the Northern Land Use Information Series,* Kananaskis Centre for Environmental Research, February 28 - March 1, 1979.

Kerr, A. (1981) Lands Directorate, Pacific Region, Vancouver (personal communication, January 20).

Kiil, A.D., Lieskovsky, R.J. and Grigel, J.E. (1973) *Fire Hazard Classification for Prince Albert National Park,* Edmonton: Environment Canada.

Krebs, C.J. and Wingate, I. (1976) "Small Mammal Communities of the Kluane Region: Yukon Territory", *Canadian Field Naturalist,* 90:379-389.

Lacate, D.S. (1969) *Guidelines for Biophysical Land Classification,* Department of Fisheries and Forestry, Canadian Forestry Service Publication No. 1624, Ottawa: Queen's Printer.

Lacate, D.S. and Romaine, M.J. (1978) "Canada's Land Capability Program", *Journal of Forestry,* 76(10):669-671.

Larsen, D.L. (1980) *Mountain Caribou Movements in the Squanga Lake Area: A Progress Report* Whitehorse: Yukon Wildlife Branch.

Larsen, D.L. (1981) Yukon Wildlife Branch, Whitehorse (personal communication, March 17).

Lavigne, D.M., Oritsland, N.A. and Falconer, A. (1977) *Remote Sensing in Ecosystem Management,* Oslo: Norsk Polarinstitutt.

Lavkulich, L.M. (1973) *Soils-Vegetation-Landforms of the Wrigley Area, North West Territories,* Task Force on Northern Oil Development Report, No. 73-18, Ottawa: Information Canada.

Lewis, P.H., Jr. (1964) *The Outdoor Recreation Plan,* Madison, Wisconsin: Wisconsin Department of Resources Development.

Lewis, P.H., Jr. (1965) "Parks and Recreation in Minneapolis", *Environmental Design Concepts for Open Space Planning in Minneapolis and its Environs, Vol. 3,* Minneapolis: Board of Parks Commissioners.

Lopoukhine, N., Prout, N.A. and Hirvonen, H.E. (1978) *The Ecological Land Classification of Labrador: A Reconnaissance,* Ottawa: Environment Canada, Lands Directorate.

Mackay, J.R. (1970) "Disturbances to the Tundra and Forest Tundra Environment of the Western Arctic", *Canadian Geotechnical Journal,* 7:420-32.

Margules, C. and Usher, M.B. (1981) "Criteria Used in Assessing Wildlife Conservation Potential: A Review", *Biological Conservation* 21:79-109.

Mitchell, B. (1979) *Geography and Resource Analysis,* London: Longman.

Murie, A. (1962) *The Wolves of Mount McKinley,* San Francisco: Mount McKinley Natural History Association.

Murie, A. (1963) *Birds of Mount McKinley Alaska,* San Francisco: Mount McKinley Natural History Association.

Murie, A. (1971) *The Wolves of Mount Mckinley,* Fauna of the National Parks of the United States Fauna Series, No. 5, Washington: U.S. Government Printing Office.

Naysmith, J.K. (1971) *Canada North-Man and the Land,* Ottawa: Information Canada.

Nelson, D., Harris, G.A. and Hamilton, T.E. (1978) "Land and Resource Classification: Who Cares?" *Journal of Forestry,* 76(10):644-649.

Nelson, J.G. (1980) "Northern Pipelines and Environmental Strategies: Some Reflections on a Management Assessment Model", Prepared for the Symposium on Environmental Management Strategies: Past, Present and Future, The Alberta Society of Professional Biologists and Alberta Environment, Edmonton, Alberta.

Novakowski, N.S. (1970) *Fire Priority Report Wood Buffalo National Park,* Prepared for Canadian Wildlife Service.

Oke, T. (1978) *Boundary Layer Climates,* London: Methuen; New York: Wiley.

Oswald, E. (1980) Canadian Forestry Service, Pacific Forest Research Centre, Victoria (personal communication, July 7).

Oswald, E.T. and Senyk, J.P. (1977) *Ecoregions of Yukon Territory,* Victoria: Fisheries and Environment Canada, Canadian Forestry Service.

Parks Canada (1973) *Environmental Analysis: A Review of Selected Techniques,* Ottawa: Parks Canada.

Parks Canada (1980) *Natural Resource Management Process Manual,* Ottawa: Parks Canada.

Pearson, A.M. (1975) *The Northern Interior Grizzly Bear Ursos Arctos L.,* Canadian Wildlife Service Report, Series No. 34, Ottawa: Information Canada.

Peterson, R.T. (1969) *A Field Guide to Western Birds,* Boston: Houghton Mifflin Co.

Pierce, T.W. and Thie, J. (1976) "Biophysical Land Classification in the Environmental Management Service", in *Ecological (Biophysical) Land Classification in Canada,* Ecological Land Classification Series, No. 1, Thie, J. and Ironside, G. (eds.), pp.55-58, Ottawa: Environment Canada, Lands Directorate.

Rogers, P.M. (1974) *Land Classification and Patterns of Animal Distributions in the Management of National Parks: Coto Donana, Spain,* Guelph: University of Guelph, unpublished M.Sc. thesis.

Rostad, H.P.W., Kozak, L.M. and Acton, D.F. (1977) *Soil Survey and Land Evaluation of the Yukon Territory,* Saskatoon: Saskatchewan Institute of Pedology.

Rostad, H.P.W., Kozak, L.M. and Acton, D.F. (1979) "User Orientated Interpretations of Soil Surveys in Yukon and Northwest Territories", in *Applications of Ecological (Biophysical) Land Classification in Canada,* Ecological Land Classification Series, No. 7, Rubec, C.D.A. (ed.), pp.345-50, Ottawa: Environment Canada, Lands Directorate.

Rowe, J.S. (1961) "The Level of Integration Concept and Ecology", *Ecology,* 42(2):420-27.

Rowe, J.S. (1972) "Fire in Northern Canada", in *Hydrologic and Biologic Studies Related to Land Use Near Watson Lake, Yukon Territory",* Murray, J.M. (ed.), pp.136-137, Ottawa: Department of Indian and Northern Affairs.

Rowe, J.S. (1979) "Revised Working Paper on Methodology/Philosophy of Ecological Land Classification in Canada", in *Applications of Ecological (Biophysical) Land Classification in Canada,* Ecological Land Classification Series, No. 7, Rubec, C.D.A. (ed.), pp.23-30, Ottawa: Environment Canada, Lands Directorate.

Rowe, J.S. (1980) "The Common Denominator of Land Classification in Canada: An Ecological Approach to Mapping", *The Forestry Chronicle,* (February): 19-20.

Rykiel, E.J. (1979) "Ecological Disturbances", in *The Mitigation Symposium,* A National Workshop on Mitigating Losses of Fish and Wildlife Habitats, Colorado State University, July 16-20, 1979, pp.624-626, Fort Collins, Colorado: U.S. Department of Agriculture, Forest Service.

Sargent, F.O. and Brandes, J.H. (1976) "Classifying and Evaluating Unique Natural Areas for Planning Purposes", *Journal of Soil and Water Conservation,* May-June.

Sauchyn, D.J. and Bastedo, J.D. (1980) "Project Carcross: A Process-Oriented Classification of the Natural Environment Surrounding Carcross, Yukon", Waterloo: University of Waterloo, unpublished typescript.

Sigafoos, R.S. (1952) "Frost Action as a Primary Physical Factor in Tundra Plant Communities", *Ecology,* 33(4):480-487.

Smith, B. (1980) Yukon Wildlife Branch (personal communication, September 16).

Steiner, F. and Brooks, K. (1980) *Ecological Planning: A Review,* Scientific Paper No. 5816, College of Agriculture Research Center, Washington State University.

Steinitz, C., Murray, T., Stinton, D. and Way, D. (1969) *A Comparative Study of Resource Analysis Methods,* Boston: Harvard University, Department of Landscape Architecture.

Tarnocai, C. (1979) "Soil Resource Inventories: Their Methods, Approaches, Interpretations", in *Resource Inventory Workshop*, Proceedings, Whitehorse, October 16-18, 1979, pp.111-118. Whitehorse: Government of the Yukon Territory.

Terry, R.D. and Chilingar, G.V. (1955) "Comparison Charts for Visual Estimation of Foliage Cover", *Journal of Sedimentary Petrology*, 25(3):229-234.

Theberge, J.B. (1972) *Considerations for Fire Management in Kluane National Park*, Report prepared for the Canadian Wildlife Service.

Theberge, J.B. (1974) *Survey of Breeding Bird Abundance, Kluane National Park*, Report prepared for Canadian Wildlife Service.

Theberge, J.B. (1976) "Bird Populations in the Kluane Mountains, Southwest Yukon, with Special Reference to Vegetation and Fire", *Canadian Journal of Zoology*, 54(8):1346-1356.

Theberge, J.B., Nelson, J.G. and Fenge, T. (eds.) (1980) *Environmentally Significant Areas of the Yukon Territory*, Ottawa: Canadian Arctic Resources Committee.

Theberge, J.B. and Fitzsimmons, M. (in preparation) *Resource Survey of the MacMillan Pass Environmentally Significant Area, Yukon: An Evaluation of Biotic Resources*, and *Resource Survey of the Kluane North Environmentally Significant Area, Yukon: An Evaluation of Biotic Resources*, Waterloo: University of Waterloo.

Thie, J. (1974a) "Remote Sensing for Northern Surveys and Environmental Monitoring", in *Canada's Northlands, Ecological Land Classification Series*, No. 0, Romaine, M.J. and Ironside, G.R. (eds.), pp.31-35, Ottawa: Environment Canada, Lands Directorate.

Thie, J. (1974b) "Remote Sensing for Northern Inventories and Environmental Monitoring", Discussion paper prepared for National Workshop to Develop and Integrated Approach to Northern Baseline Data Inventories, April 17-19, 1974, Toronto.

Thomas, J.W. (ed.) (1979) *Wildlife Habitats in Managed Forests: The Blue Mountains of Oregon and Washington*, Washington: U.S. Department of Agriculture, Forest Service.

Truett, J.C. (1979) "Pre-impact Process Analysis-Design for Mitigation", in *The Mitigation Symposium: A National Workshop on Mitigating Losses of Fish and Wildlife Habitats*, July 16-20, 1979, pp. 355-360, Fort Collins, Colorado, U.S: Colorado State University, Department of Agriculture, Forest Service.

Van de Ploeg, S.W.F. and Vlijm, L. (1978) "Ecological Evaluation, Nature Conservation and Land Use Planning with Particular Reference to Methods Used in the Netherlands", *Biological Conservation*, 14:197-219.

Walmsley, M.E. and Van Barneveld, J. (1979) "Biophysical Land Classification Techniques: The Application to Ecological Classification of Land in British Columbia", in *Resource Inventory Workshop*, Proceedings, Whitehorse, October 16-18, 1979, pp.7-32, Whitehorse: Government of the Yukon Territory.

Walmsley, M.E., Utzig, G., Vold, T., Moon, D. and van Barneveld, J. (eds.) (1980) *Describing Ecosystems in the Field,* RAB Technical Paper 2, Victoria: British Columbia, Ministry of Environment.

Webber, P.J. and Ives, J.D. (1978) "Damage and Recovery of Tundra Vegetation", *Environmental Conservation,* 5(3):171-82.

Webster, N. (1969) *Webster's New Twentieth Century Dictionary of the English Language,* New York: The World Publishing Company.

Wiken, E.B. and Ironside, G.R. (eds.) (1977) *Ecological (Biophysical) Land Classification in Urban Areas,* Ecological Land Classification Series, No. 3, Ottawa: Environment Canada, Lands Directorate.

Wiken, E.B. and Welch, D. (1979) "An Interdisciplinary Basis for Resource Studies: The Ecological Land Survey", in *Resource Inventory Workshop,* Proceedings, Whitehorse, October 16-18, 1979, pp.33-5, Whitehorse, Government of the Yukon.

Wiken, E.B. and Welch, D.M. (1980) "An Interdisciplinary Basis for Resource Studies: The Ecological Land Survey", in *Resource Inventory Workshop,* Proceedings, Whitehorse, October 16-18, 1979, pp.33-50, Whitehorse: Government of the Yukon Territory.

Wiken, E.B., Welch, D.M., Ironside, G. and Taylor, D.G. (1978) "Ecological Land Survey of the Northern Yukon", *Applications of Ecological (Biophysical) Land Classification Series,* No. 7, Rubec, C.D.A. (ed.), pp.361-372, Ottawa: Environment Canada, Lands Directorate.

Willard, B.E. and Marr, J.W. (1971) "Recovery of Alpine Tundra Under Protection After Damage by Human Activities in the Rocky Mountains of Colorado", *Biological Conservation,* 3(3):181-90.

Wratten, S.D. and Fry, G.L.A. (1980) *Field and Laboratory Exercises in Ecology,* Baltimore: University Park Press.

Wright, D.F. (1977) "A Site Evaluation Scheme for Use in the Assessment of Potential Nature Reserves", *Biological Conservation,* 11:293-305.

Yukon, Government of (1979) *Resource Inventory Workshop,* Proceedings, Whitehorse, October 16-18, 1979, pp. 7-32, Whitehorse: Government of the Yukon Territory.

Yukon, Government of (1980) *Yukon Land Resources and Inventory Atlas,* Whitehorse: Government of the Yukon.

Zoltai, S.C. (1979) "Ecological Land Classification Projects in Northern Canada and Their Use in Decision Making", in *Applications of Ecological (Biophysical) Land Classification in Canada,* Ecological Land Classification Series, No. 7, Rubec, C.D.A. (ed.), pp.373-81, Ottawa: Environment Canada, Lands Directorate.

Zoltai, S.C. (1981) Lands Directorate, Ottawa (personal communication, February 10).

APPENDIX A

EXCERPTS FROM ENVIRONMENTALLY SIGNIFICANT AREAS OF THE YUKON REPORT (THEBERGE, NELSON AND FENGE, 1980)

It is imperative that representative, unique, and sensitive lands be identified and managed so that they may continue to function, as far as possible, as natural self-regulating ecosystems (p.4).

[There is a] need to set aside some wild, relatively unaltered areas for wildlife and renewable resource protection as well as for enjoyment, study, and safe-keeping (p.4).

[A] commitment [is needed] to protect key wildlife and natural areas . . . (p.4).

The majority of areas suggested as ESAs are thought to be incapable of withstanding multiple use, but detailed site research will determine this (p.20).

These development activities [exploitation of minerals and other resources] should be conducted primarily on lands that are not designated as ESAs (p.1).

The objective of this [first phase] inventory was to identify the bio-physical features, functions, and processes which justify the preservation/protection of these lands and the land use and management implications of these findings (p.12).

APPENDIX B

CONSIDERATIONS FOR RESOURCE SURVEYS OF ESAS IN NORTHERN ENVIRONMENTS

The selection or design of a resource survey method for a particular project must be considered not only within the context of pre-specified terms of reference but also within the context of the environmental setting in which the survey is to be undertaken; different environments may require different types of resource surveys. The major factor determining differences among various environments is climate. Light, heat, moisture and air flow regimes peculiar to an area directly or indirectly influence the properties of most ecological processes, including human activities. These properties in turn influence the type, quality, quantity, distribution and use of regional and local resources. It follows that, to be suitable for application to a given area, a resource survey method must account for climate-induced attributes which distinguish that area from others.

In this appendix, distinctive attributes of the environment of Canada's north (Yukon and Northwest Territories) are discussed in relation to the climate of northern latitudes. Effects of *macroclimate* (regional climate factors) and *microclimate* (local horizontal and vertical climate factors) are dealt with separately as they influence the environment in different but complementary ways. Implications of these effects on the design of a resource survey for the Yukon's ESAs are then discussed.

Macroclimatic Effects on the Northern Environment

In northern Canada, as elsewhere, the amount of solar radiation received by the land surface is the most important factor determining regional weather patterns, especially temperature. While there is some modifying influence of macroclimate from the Arctic Ocean (Oswald and Senyk, 1977), it is the relatively low levels of radiant input which create the north's characteristically cooler climate.

Abiotic Component

From an abiotic standpoint, the most significant consequence of cooler temperatures in the north is the occurrence of permafrost, a layer of perennially frozen soil. Associated with permafrost are conditions of poor drainage which, in flat areas, result in distinctive assemblages of lake and wetland ecosystems. As well, the alternate freezing and thawing of soil layers above the permafrost level create symmetrically patterned land forms and microrelief features unique to northern environments. In some areas this process is the most important factor influencing the development and composition of plant communities (Sigafoos, 1952). Disturbance of these communities by fire or human activities may result in partial thawing of the permafrost layer (due to loss of insulating vegetation) followed by incremental erosion or subsidence of large volumes of soil (Mackay, 1970). These macroclimate/abiotic effects suggest the following resource survey guidelines:[29]

1. Map the extent, character and relative intensity of frost action.
2. Include considerations of frost action in classification scheme, especially in relation to abiotic-vegetation associations.
3. Identify representative ecosystems or land forms created by frost action for purposes of research and/or interpretation.

Biotic Component

Within the biotic component of northern environments, reduced thermal input is reflected primarily in the low productivity levels among plant and animal communities. Dickinson (1978) summarized this effect:

> With large areas of low plant productivity, much land is required to sustain populations of the larger herbivores and carnivores. The timing of reproduction in many northern animals, both migratory and resident, is critical and has a very narrow margin of safety. Consequently, there is great fluctuation in reproductive success or survival of young from year to year. In addition, the North is characterized by biotic communities with a low diversity of species. Food chains, therefore, are relatively short, and fluctuations in numbers of one species may profoundly affect other species. Cycles, such as those characteristic of snowshoe hare and lynx, are common in the North. Slow growth rates are characteristic of other animals, such as marine [and freshwater] invertebrates and fish. (Dickinson, 1978, p.302)

----

[29] These and subsequent guidelines in this chapter are proposed without regard to economic or logistical constraints.

The presence of boreal vegetation communities, found to a varying extent throughout the north, is directly related to macroclimatic factors related to temperature.[30] These communities, consisting largely of cold-adapted coniferous species, are subject to frequent and random burnings as indicated by the typical mosaic pattern created by adjacent vegetation communities showing various stages of post-fire growth. In general, fires have a favourable effect on biological systems by restoring nutrients to the soil, renewing vegetation and increasing the diversity of habitats available to wildlife.

With respect to reduced plant productivity, it is worth mentioning that some tundra communities may represent the outcome of prolonged periods of growth. Willard and Marr describe such communities in relation to the estimated time periods needed for re-growth following disturbance:

> The time-factor in tundra recovery is quite shocking. We estimate that some tundra . . . damaged by only a few seasons of human activity will require hundreds of years, possibly even a thousand, to rebuild a natural and persistent . . . ecosystem. (Willard and Marr, 1971, pp.181-221)

In areas of considerable frost action, the chances for total recovery of disturbed communities are reduced due to edaphic changes following erosion or subsidence.

These macroclimate/biotic effects suggest the following resource survey guidelines:

1. Develop classification criteria and data acquisition procedures which are applicable to both sedentary and wide ranging wildlife species.
2. Incorporate temporal as well as spatial considerations in identifying critical wildlife areas.
3. Map zones within which seasonal movements of ungulates have been regularly recorded.
4. Identify vegetation communities of low recoverability due to frost action or advanced age.
5. Map fire susceptibility of vegetation communities for purposes of wildlife habitat management and protection of human populations.

Cultural Component

As macroclimatic factors have created uniquely northern characteristics among biological systems, so have these systems influenced cultural processes related to the perception and use of resources. Dickinson (1978) illustrates this link by relating low biological productivity to the traditional perception

----

[30] The word "boreal" etymologically means "northern".

of wildlife as a key northern resource:

> Native peoples of the North have evolved their ways of living in relation to . . . relatively low levels of biological productivity; and of necessity they have been mainly dependent on wild animals, rather than on wild plants . . . While this dependence on wild animals has been reduced with the introduction of imported foods, it is still of considerable importance in many regions. (Dickinson, 1978, p.302)

Before the importation of food resources, the availablity of wildlife was a major factor in controlling the size of human populations (Crowe, 1974). Even as imported foods became more common, it was largely a combination of inhospitable climatic factors and limited access which kept the size of resident populations in check and discouraged immigration by non-natives. During the early 1900s, these controls were substantially overcome in some areas with the importation of fuel energy and advancements in the technology associated with transportation.

Over the past eighty years, southern interests in the exploitation of non-renewable resources such as minerals and fossil fuels have resulted in a rapid development of transportation systems - roads, trails, airports, and, in the Yukon, a major rail route - with an associated net immigration rate among the highest in Canada's history (Dickinson, 1978). Corresponding to the relatively rapid growth in human populations has been the proliferation of such cultural processes as settlement, recreation, forestry, mining, resource exploration and hunting.

All of these activities have, in general, been undertaken without prior knowledge of their effects on abiotic and biotic components of the environment, due primarily to the long time frame required for scientific research in northern climates. Dickinson (1978) summarizes these deficiencies in northern knowledge:

> There are basically two levels of knowledge gaps; first, how do biotic communities [and abiotic systems] within northern ecosystems function when man plays a relatively small part within those communities [and systems]; and second, when man plays a significant part, what will be the effects of his actions? It is not possible to answer questions of the second category adequately without reference to the first. Nevertheless . . . research [has addressed] only questions of the second category, . . . [consisting of the collection] of the simplest baseline data in the shortest possible time. (Dickinson, 1978, p.307)

These macroclimate/cultural effects suggest the following resource survey guidelines:

1. Identify areas used traditionally by natives for hunting and for gathering of food plant species.

2. Map sequences of land use change through time with special reference to transportation, mining and resource exploration.
3. Map types and intensity of hunting pressures through time.
4. Identify areas offering opportunity to monitor long term effects of past impacts caused by man.
5. Identify unique and/or representative ecosystems offering opportunity for long-term baseline studies.

Microclimatic Effects on the Northern Environment

Superimposed upon the distinctly northern macroclimate are vertical and horizontal zonations among climate factors caused by variations in radiant input due to local topography. In general, the effect of moving from low to high latitudes is to decrease the illumination of north-facing slopes in favour of south-facing slopes (Oke, 1978). At middle latitudes (41 degrees) south-facing slopes may receive almost three times more net radiation than comparable north-facing slopes (Oke, 1978). At high latitudes (60+ degrees) the effect of topographic slope changes is increased so that small slope or aspect differences may be of considerable ecological significance (Oke, 1978).

Abiotic Component

Within the abiotic component, topographically induced variations in the temperature and moisture directly or indirectly govern the rate, magnitude and periodicity of such geomorphic processes as frost shattering, nivation, mud slumping, soil creep, rock exfoliation, landslides, debris flows and avalanches.

Oswald and Senyk describe the relationship between microclimatic factors and permafrost:

> Variations in local terrain conditions are responsible for the occurrence of permafrost in the discontinuous zone and variations in thickness of the active layer in both the continuous and the discontinuous zones. (Oswald and Senyk, 1977, p.19)

These microclimate/abiotic effects suggest the following resource survey guidelines:

1. Map hazardous areas related to landslides, debris flows or avalanches with special reference to the relative rate, magnitude and periodicity of these processes.
2. Identify areas especially susceptible to thermokarst erosion and subsidence.

Biotic Component

Variations in microclimate strongly influence the relative productivity of biological systems. In mountainous or hilly terrain, the concentration of plant biomass in valley areas is largely a response to a more favourable microclimate than is found at higher, more exposed elevations. Protected valleys north of the latitudinal treeline are heavily vegetated for similar reasons. Many species of wildlife also concentrate in these areas for food, migration and/or thermal protection. Due to the availability of fuel, fire is most common on these sites, contributing to their already high levels of productivity (from regeneration) and creating ecotone and habitat for wildlife.

Variations in thermal and air flow regimes on slopes of different angle and aspect may result in distinctly different plant associations. For example, a pattern common in hummocky lowland areas or on moderate sub-alpine slopes is the presence of xeric grassland communities on southern aspects in contrast to forested northern aspects. The species composition and/or sequence of seral stages is also strongly influenced by differences in aspect. Such variations in species composition and successional trends are reflected in the different wildlife assemblages using different aspects.

These microclimate/biotic effects suggest the following resource survey guidelines:

1. Develop classification scheme which highlights areas of relatively high biological productivity.
2. Include consideration of microclimatic variants during interpretations related to succession or potential vegetation.
3. Identify unique plant communities or wildlife habitats which are an artifact of microclimatic variations.

Cultural Component

Microclimatic factors in the north have strongly influenced the distribution of such land uses as hunting, settlement, forestry and transportation. Historically, native populations were attracted to areas of high biological productivity namely valleys and other protected sites, for purposes of hunting, shelter and fuelwood supply. Present patterns of settlement and transportation have developed largely in response to this natural tendency. The resultant *corridor effect* is typical of land use patterns in northern environments. Related to this cultural phenomenon is a higher frequency of man-caused forest fires in valley areas.

These microclimate/cultural effects suggest the following resource survey guidelines:

1. Develop classification scheme which highlights the corridor effect in land use.
2. Concentrate field sampling in valleys and other protected areas.

3. Map past forest fires in relation to their age and probable cause - natural or man-caused.
4. Identify forests highly susceptible to fire which are in close proximity to present or potential human settlement.

APPENDIX C

YUKON INFORMATION SOURCES FOR LITERATURE SEARCH AND INTERVIEWS

Among the first references to be consulted should be the *Yukon Land and Resource Inventory* (Yukon, 1979) and the appropriate *Land Use Information Series* (Canada, 1972-1981) map. Earlier surveys, scientific research and historical data may be found in the Yukon Public Archives in Whitehorse. For interviews, as well as additional literature, the following agencies in Whitehorse should be contacted: Canadian Wildlife Service, (DOE);[31] Yukon Wildlife Branch, (YTG);[32] Fisheries Service, (DOE); Yukon Forestry Service, (YTG); Canadian Forestry Service, (DIAND);[33] National and Historics Parks Branch, (DOE); Tourism and Information Branch, Land Use Office, (YTG); Mining Recorder's Office, (DIAND), and the Council of Yukon Indians.

---

[31] Department of Environment.
[32] Yukon Territorial Government.
[33] Department of Indian Affairs and Northern Development.

APPENDIX D

FIELDSHEET FOR BIOTIC LAND UNITS

FIELDSHEET FOR BIOTIC LAND UNITS

Attached notes on section(s): ______________________________

1. ESA ______________________________
2. Date ______________________________
3. Investigators ______________________________
4. BLU ______________________________
5. NTS sheet ______________________________
6. UTM grid no. ______________________________
7. Site Description
   a) elevation ______________________________
   b) position (macro) ______________________________
   c) relief shape ______________________________
   d) slope ______________________________
   e) aspect ______________________________
   f) exposure type ______________________________
   g) moisutre regime ______________________________
8. Major Vegetation Community
   a) areal coverage within BLU (%) ______________________________
   b) physiognomic class ______________________________
   c) vegetation type ______________________________
   d) composition and cover (see table)
   e) tree canopy age: 0-50 ______________________________
      50-100 ______________________________
      100-150 ______________________________
      150-200 ______________________________
      200 + ______________________________
   f) fire periodicity ______________________________
   g) natural processes influencing vegetation ______________________________

8. d) composition and cover

STRATA	SPECIES	% COVER
Trees (> 5 m)		
Trees (< 5 m)		
Tall Shrubs (1.5 to 5 m)		
Medium Shurbs (0.5 to 1.5 m)		
Low Shrubs (0.1 to 0.5 m)		
Ground Shrubs (< 0.1 m)		
Forbs		
Graminoids		

STRATA	GROUP	% COVER
Ferns and Fern Allies		
Bryophytes		
Lichens		
Organic Litter		
Surface Substrate		

Figure 10 Fieldsheet for Biotic Land Units

9. Minor Vegetation Communities Within BLU

PHYSIOGNOMIC CLASS	VEGETATION TYPE	AREAL COVERAGE (%)

10. Sparsely or Non-Vegetated Areas Within BLU

PHYSIOGNOMIC CLASS	VEGETATION TYPE	AREAL COVERAGE (%)

11. General Comments on Vegetation

12. Wildlife

a) indicators of occurrence

				OTHER
SPECIES				
DROPPINGS				
TRACKS				
SIGHTINGS				
HEARINGS				
NEST/DEN				
FEEDING EVIDENCE				
OTHER				

b) significant habitat features

				OTHER
ESCAPE TERRAIN				
FORAGE SPECIES				
NEST/DEN SITES				
ECOTONE				
SHELTER/ COVER				
SALT LICK				
OTHER				

c) restricted avian habitat

d) general comments on wildlife

13. Human-Related Processes Influencing Vegetation and/or Wildlife Within BLU: (code - past (-); present (o); potential (+)

14. Photos

APPENDIX E

GUIDE TO THE FIELDSHEET FOR BIOTIC LAND UNITS

Preliminary Instructions: No items on the fieldsheet should be left blank unless the surveyor intends to add the information later (examples would be map co-ordinates or classification names). A cross (x) should be put through the item number when the item is not sampled or when the surveyor is unsure as to the correct value. A slash (/) should be put through the item number when the item is looked for but not found (an example would be evidence of fire periodicity in a treeless alpine environment).

1. *ESA*

   The name and/or number of the ESA under study should be noted for each sample area.

2. *Date*

   The date of sampling is recorded simply as year/month/day. For example 81/06/07 translates as June 7, 1981.

3. *Investigators*

   Initials of those involved in the sampling are sufficient.

4. *BLU*

   The biotic land unit (BLU) under study is recorded in code form according to its Landmark Area and map unit number.

5. *NTS Sheet*

   Space is provided for the name, number and scale of the National Topographic System (NTS) mapsheet within which the BLU is being sampled. To aid future orientation, the largest available map scale is preferable.

6. *UTM Grid No.*

As a fixed geographic reference for locating the sampling area, the Universal Trans Mercator (UTM) grid number associated with the area is recorded from the above mapsheet.

7. *Site Description*

To facilitate vegetation inferences to unsampled units, edaphic associations related to major community are described including elevation, site position (macro), relief shape, slope aspect, exposure type and ecological moisture regime (Appendix F).

8. *Major Vegetation Community*

This section of the fieldsheet is designed to capture the major diagnostic features of the plant community of greatest areal coverage within the biotic land unit.

(a) *Areal Coverage within BLU (%)*

The areal coverage of the major community relative to the total area of minor communities within the biotic land unit is recorded as a per cent. This measurement can be estimated from aerial photographs, from an overhead flight (while airborne or from photos taken in craft) or from a nearby high vantage point.

(b) *Physiognomic Class*

A physiognomic class is assigned to the major plant community which describes its salient structural features. The appropriate alpha-numeric classification (Table 6c) is assigned on the basis of the type and per cent cover of the community's dominant stratum. The classification system includes six easily recognizable community classes - forest, woodland, shrubland, meadow, bryoid mat[34] and wetland - and their subclasses, plus two classes for sparsely and non-vegetated areas. Designed to accommodate all plant communities found in the Yukon, the classification system reflects inputs from a number of biologists familiar with the territory.

----

[34] In designing this system, the author purposely avoided the term *tundra* (Webber and Ives, 1978) as it embraces several types of plant communities which are physiognomically distinct. The *bryoid mat* class eliminates the possiblity of misclassification and applies primarily to alpine or arctic communities (see Table 6c).

(c) *Vegetation Type*

This descriptor provides a simple reference to the major community with respect to the plant species which comprise the greatest amount of per cent cover within the sampling area. The appropriate codes for major species are recorded (first three letters of the plant's genus and species) and joined with a hyphen.

(d) *Composition and Cover*

This section of the fieldsheet is used to record the species code or group and per cent cover of plants within each of eleven height classes or strata. Two other classes allow for the description of deadfall and non-vegetated surfaces. To arrive at consistent use of this section, the following definitions and directives are discussed.

Per Cent Cover is the per cent of the ground area covered by each plant species, plant group, or type of organic litter, or surface substrate. In the case of woody plants, care must be taken not to bias estimates towards foliage density; in all cases the area within the perimeter of a vertical projection of the plant onto the ground is assumed to be covered. Because of the overlaying of different strata, the summed cover value will be in excess of 100 per cent in all but sparsely vegetated communities (Wratten and Fry, 1980). To aid in the assessment of cover, a set of comparison charts is presented in Appendix G.

The Strata:

Trees have been divided into two strata: those greater and less than 5 metres in height. In the Yukon, typical trees include alpine fir, white and black spruce, lodgepole pine, trembling aspen, balsam poplar and paper birch.

Shrubs are plants with persistent woody stems and a relatively low growth form. They usually produce several basal shoots as opposed to a single bole as in trees (Thomas, 1979). Tall shrubs typically include various species of willow; medium shrubs include willows and mountain alder; and low shrubs may include shrub birch, labrador tea, soapberry *(Shepherdia canadensis)*, *Ribes* spp., *Rubus* spp. and *Vaccinium* spp. Ground shrubs are prostrate woody plants which commonly include *Arctostaphylos* spp., *Dryas* spp., crowberry *(Empetrim nigrum)*, twinflower *(Linnaea borealis)*, heathers *(Cassiope* spp., *Phyllodoce* spp.) and trailing willows *(Salix reticulata, S. polaris, S. arctica)*.

Forbs include any herbaceous plant species other than those in the Graminae, Cyperaceae and Juncaceae families (Thomas, 1979). The species code and per cent cover of up to five forbs should be recorded regardless of their height. Where identification in the field is at all in doubt, specimens should be sampled and pressed for later reference. Other forbs which occur infrequently and may be relatively rare should also be collected as vouchers.

Graminoids include all herbaceous plants other than forbs. Because of the relative difficulty in identifying these plants, this stratum should be sampled more intensively than forbs. The field person should aim to recognize graminoids on site at least to the level of genus.

Ferns and Fern Allies collectively refers to all ferns, club mosses (*Lycopodium* spp.), spike mosses (*Selaginella* spp.) and horsetails (*Equisetum* spp.). These plants need not be identified down to species; rather they should be grouped according to genus.

Bryophytes refer to all ground mosses. Due to the difficulties in identifying moss species, they have been put into six species groups based on commonly occurring associations which can be recognized in the field with relative ease. These groups, developed by Oswald (1980), are shown in Appendix H. Ground *Lichens* have been grouped similarly and are shown in the same appendix. In both cases the name and per cent cover of the group(s) should be recorded in the space provided.

Organic Litter includes all dead and down woody material such as logs and limb piles, snags, plus leaf, cone and needle litter. Significant both as fuel for fires and as cover for various wildlife, this material is recorded as to its group (Appendix H) and its per cent cover is estimated as for plant cover.

Surface Substrate such as exposed bedrock, boulders, cobbles and stones, gravel, sand, exposed mineral or peaty soil and open water is recorded to complete the description of the major community (Appendix H).

(e) *Tree Canopy Age*

An age estimate of the dominant tree canopy (if present) is obtained by coring several trees of the main tree stratum. An increment borer should be applied to at least three of the largest and healthiest trees within the sample area. For accurate readings, the sample core must lie within three years of the central pith. In selecting trees for coring, care should be taken to avoid "veterans", trees significantly older than the trees of the main canopy which have survived one or more fires. Veterans are usually isolated in distribution and often extend well above the main tree stratum (Walmsley et al., 1980).

(f) *Fire Periodicity*

The age of past burns, recorded as scars in the growth rings of surviving trees, is used as an indicator of the fire periodicity within different types of forests. Fire scars may be recognized as dark lines within the sample cores obtained when determining canopy age or as visible scars, usually blackened and confined to the basal portion of the trunk. In the latter instance, core samples should be taken from the scar site to the center of the tree to determine the number of years elapsed. The average age of the burn(s) should be recorded in the space provided. Fire history data gained from tree disc analysis as well as literature research or interviews should contribute to this estimation.

(g) *Natural Processes Influencing Vegetation*

The objective of this section of the fieldsheet is to identify, through observable or reported evidence, historical and/or ongoing natural processes relevant to the appearance of present vegetative cover within the sample area. Presented below is a list of processes which may influence vegetation. It is not possible to compile a list which encompasses the myriad of factors which may affect local site conditions; therefore, additional categories may need to be added. The codes listed beside each category should be used when recording the information. Except for avalanches (A1) and beaver dams/ponds (W3), all of the abiotic, fire and wildlife-related factors may be investigated within the sample area. The above exceptions should be recorded as present only if they are outstanding characteristics of the biotic land unit as a whole.

W. Wildlife-related effects

1.) Heavy browsing by ungulates (at least 20 per cent of shrubs in sampling area browsed)
2.) Tree-cutting by beavers (primarily aspen, paper birch and balsam poplar)
3.) Beaver dams/ponds (inundation)

F. Fires

1.) Intensive Fire (charred standing or fallen deadwood or charred stumps)
2.) Light Fire (primarily ground fire; singed bark or exposed roots)

A. Abiotic processes

1.) Avalanche (avalanche scars)
2.) Eolian action (active deflation or deposition)
3.) Frost action (mud boils, stripes, stone circles, palsas, polygons, solifluction, peat plateaus, tussocks, hummocks)
4.) Nivation (regular patches of alpine vegetation and lichen-less rocks)

9. *Minor Communities within BLU*

Procedures for describing minor communities ((a) physiognomic class, (b) vegetation type and (c) per cent areal coverage within BLU) are the same as for the major community. A short visit to these communities is recommended, if time is available, in order to verify descriptions and to sample unusual plants.

10. *Sparsely and Non-vegetated Areas within BLU*

The physiognomic class and type of non-vegetated areas are recorded following the classification system in Table 6c. Per cent areal coverage is estimated as for plant communities.

11. *General Comments on Vegetation*

This section allows for the recording of any interesting items of information whose discovery is unforeseen yet worth storing. Such "serendipitous findings" (Mitchell, 1979) could include rare species of plants, unusual diversity of communities, vegetation features or processes of special interest for interpretation or scientific research, etc.

12. *Wildlife*

The following sections of the fieldsheet record data related to wildlife occurrence and to an assessment of significant habitat features.

(a) *Indicators of Wildlife Use*

To provide an overview of current wildlife distribution within an ESA, faunal assemblages associated with biotic land units are identified by using a nominal checklist system in which to record the occurrence[35] of wildlife signs. To maximize predictability of results, sampling should largely be confined to the major community of a selected BLU. The entire sampling area should be systematically searched for droppings, tracks/trails and evidence of feeding and any findings noted on the fieldsheet with a checkmark. Nests, dens, bedding sites, hearings and sightings discovered outside of the sampling area, but within the major community, should also be recorded.

(b) *Significant Habitat Features*

The potential for a selected BLU to serve as habitat for a particular species is assessed by using the checklist in conjunction with the one described above. Again, most of the sampling is confined to the sampling area. However, because of their exceptional importance to ungulates, salt licks discovered anywhere within the BLU should be also recorded. Specific notes aiding the recognition of significant habitat features for mammals and birds are presented in Appendix I and J respectively. For mammals, these features aid in the recognition of preferred breeding habitat, major food sources, cover requirements and ecotone. For birds, features of their preferred breeding and/or foraging habitat are described.

(c) *Restricted Avian Habitat*

Some birds cannot be correlated to particular vegetation communities but rather are restricted to habitat situations created by: (1) the type and/or proximity of water bodies in the vicinity and their associated shoreline characteristics (e.g. Bald Eagle); (2) particular surface substrate associations (e.g. Common Nighthawk); and (3) buildings or other human-built structures (e.g. Barn Swallow). Such birds for which habitat is provided within the biotic land unit should be listed in this section.

----

[35] Measures for relative abundance were not considered due to the variability of populations over time; such results from only one field season may have little representative value (Hirsch et al., 1979).

(d) *General Comments on Wildlife*

In this section, notable information related to wildlife can be recorded which was not recorded elsewhere on the fieldsheet. Examples are the description of particularly significant habitat features in minor communities, the identification of habitats high in bird species diversity, the recording of historical data or reports on local wildlife populations, etc.

13. *Human-related Processes Influencing Vegetation and/or Wildlife within BLU.*

This is an optional section of the fieldsheet which need not be filled out if the cultural component of the resource survey is undertaken concurrently. Listed below are the major types of past (-), present (0) or proposed (+) land uses which may have direct or indirect effects on biotic resources. These effects are diverse and may influence vegetation and wildlife to different degrees. For instance the ease of access provided by roads may indirectly result in relatively high hunting pressures. In another case the activities associated with hard rock mining may directly effect the structure or growth of vegetation. Space is provided to record other human-related factors which are not listed but merit notation.

S. Scattered Habitation (trapper's cabin, squatters, cottages)
U. Urban Development (villages, towns, subdivisions)
R. Recreation (beach activities-ba, canoeing-ca, camping-c, cottaging-ct, boating-bo, viewing-v, park-p, historical-h)
A. Agriculture (cultivation-cv, grazing-gr)
F. Forestry (selective and clearcut logging)
M. Mining (excavations, mine spoils, associated diggings and trails)
T. Transportation (roads-rd, railway-rl)

14. *Photos*

For each photograph taken within a sampled BLU, the film roll and frame numbers should be recorded plus a brief description of the shot.

APPENDIX F

FIELDSHEET GUIDELINES: SITE DESCRIPTION

The terms used in the site description (section 7 of the field sheet) are noted in Table 9.

The following comments on the site position macro are based on Walmsley et al. (1980). The scale of perspective of the landscape is from the tops of mountains to the floors of the main valleys, the vertical difference usually being well in excess of 300 metres in most mountain regions. For areas where the landscape is dominantly a plain or plateau, circle "g" for "Plain".

The terms used to describe site position macro are defined below, with some typical examples of surface shape encountered. Figure 11 illustrates the terms below by showing an idealized sequence for mountainous terrain.

a. *Apex*- the upper most portion of a mountain. Surface shape is often convex.
b. *Face* - the vertical rock wall with steep exposed bedrock.
c. *Upper Slope* - the generally convex upper portion of the mountain slope immediately below the apex and, if present, the face.
d. *Middle Slope* - the area of a mountain between the upper slope and the lower slope where the general slope profile is not distinctly concave nor convex.
e. *Lower Slope* - the area toward the base of the mountain slope where the broad slope profile is generally concave.
f. *Valley Floor* - lower part of the valley system, bounded on both sides by mountain ranges, and more or less horizontal in cross section. Valley floors generally have level to moderate slopes.
g. *Plain* - the area in which gravitational forces and confinement of water bodies by mountainous topography (>300m difference between mountain tops and valley floors) are not major factors in the processes of landscape formation. Plateaus are considered as elevated plains (Holland, 1976). Plains may occur at any elevation.

SITE POSITION	ASPECT	ELEVATION Feet	Metres	MORPHOLOGICAL EXPRESSION Examples	SLOPE CLASS	EXPOSURE TYPE	ECOLOGICAL MOISTURE REGIME
Apex	None	2152	656	Plain *lake bottom*	Level 0-2%	Not Applicable	xeric
Face	N	2500	762	Apron *coalescing talus with finer sediments*	Gentle 2-9%	Wind	mesic
Upper Slope	NE	3000	915	Rolling Plain *low amplitude high frequency undulations*	Moderate 10-15%	Insolation	hygric
Middle Slope	E	3500	1067	Hummocky *kame-kettle lake complex*	Strong 16-45%	Cold Air Drainage	hydric
Lower Slope	SE	4000	1220	Ridged *kames, eskers, mine tailings*	Extreme 46-70%	Saltspray	
Valley Floor	S	4500	1372	Terraced *fluvial, lacustrine benches*	Steep > 70%	Atmospheric Toxicity	
	SW	5000	1524	Fan *talus or fluvial*		Other *eg. waterfall spray*	
	W	5500	1677			Frost	
	NW	6000	1829				
		6500	1982				
		7000	2134				
		7280	2220				

Table 9 Parameters Used to Describe Ecological Conditions of Sample Area

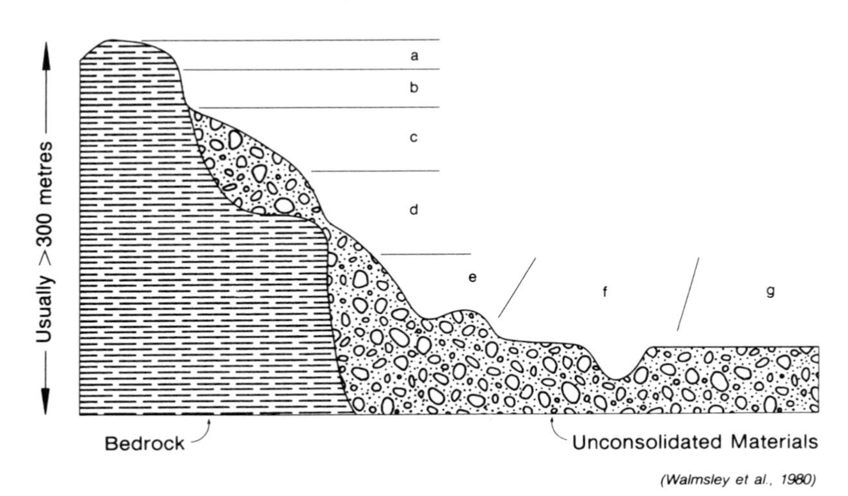

Figure 11 Schematic Cross-Sectional Diagram Illustrating Application of Terms to Describe Site Position Macro

APPENDIX G

FIELDSHEET GUIDELINES: METHODS FOR VISUAL ESTIMATION OF FOLIAGE COVER

(for less detailed vegetation data needs)

CLASS	GROUND COVERED
I	Less than 5%
II	5-25%
III	25-50%
IV	50-75%
V	75-100%

Table 10 Simplified Class System for Foliage Cover Estimation

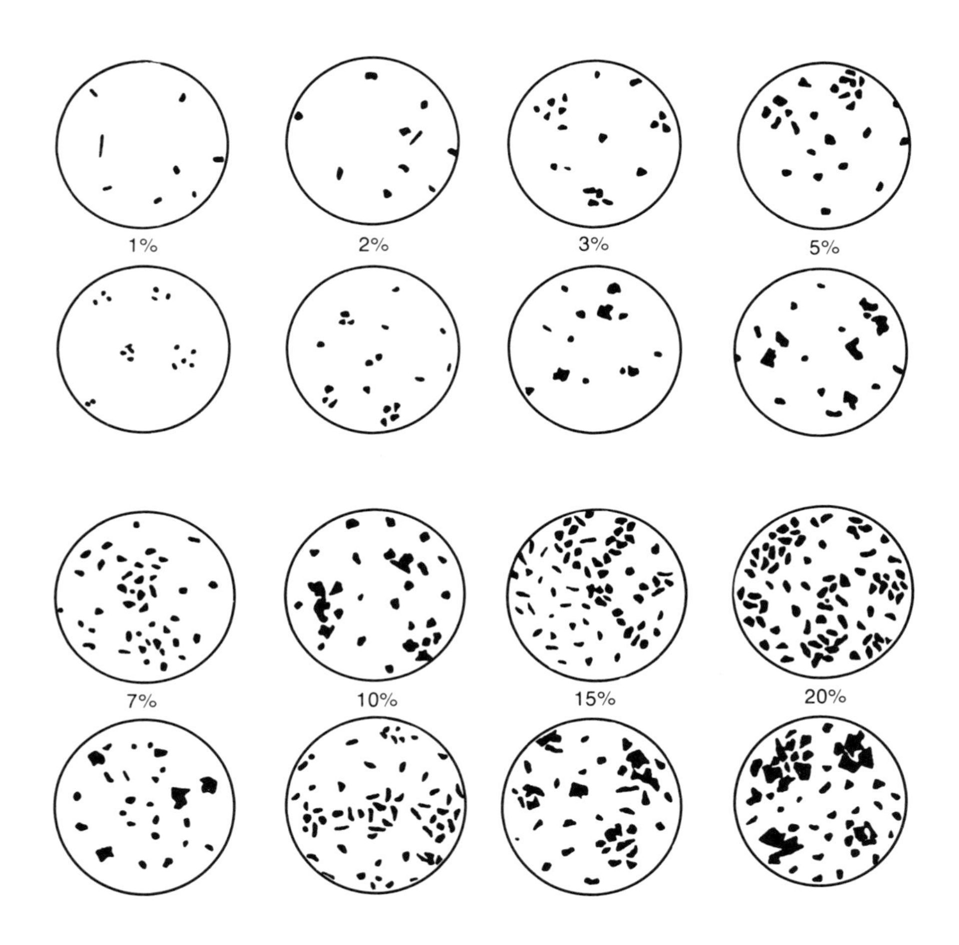

Figure 12 Foliage Cover Comparison Charts

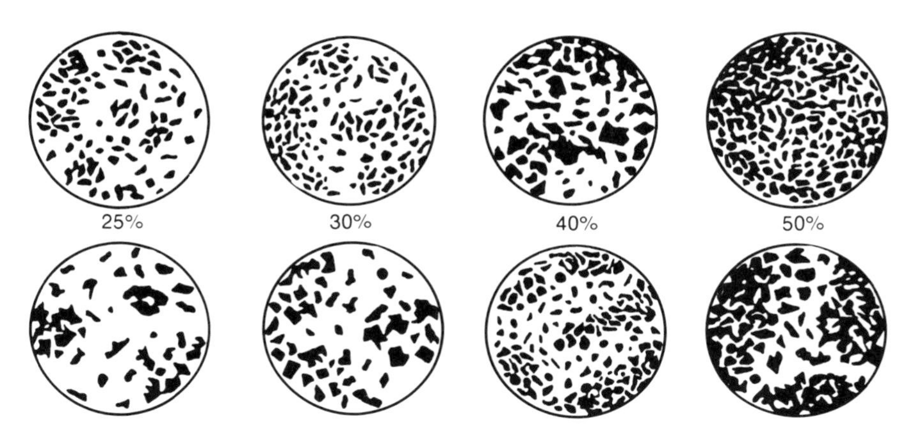

Developed by Richard D. Terry and George V. Chilingar. Published by the Society of Economic Paleontologist and Minerologists in its Journal of Sedimentary Petrology 25(3): 229-234, September 1955.

APPENDIX H

DESCRIPTION OF LOWER VEGETATION STRATA AND SUBSTRATA

STRATUM	GROUP NAME	CODE	SPECIES/MATERIAL INCLUDED
BRYOPHYTES	Feathermoss	m-fe	Hylocomium splendens Pleurozium schriberi Ptilium crista-castrensis Dicranum spp. Abietinella abietina
	Bog Moss	m-bg	Aulacomnium turgidum A. palustre Tomenthypnum nitens Sphagnum spp.
	Wetland Moss	m-wt	Drepanocladus uncinatus Fontinalis antipyratica
	Alpine Moss	m-al	Rhacomitrium canescens Dicranum spp. Grimmea spp.
	Invading Moss	m-in	Ceratodon purpureus Funaria hygromtrica
	Dryland Moss	m-dr	Rhacomitrium canascens Dicranum spp. Polytrichum juniperinum P. piliferum Tortula spp.
LICHENS	Reindeer Lichens	l-rd	Cladina mitis C. rangiferina C. alpestris
	Forest Lichens	l-fo	Cladonia genus other than Caldina subgroup, of which there are many species, e.g. C. gracilis C. pyxidata C. coccifera C. furcata C. uncialis Nephrona arcticum Peltigera rufescens P. apthosa
	Alpine Lichens	l-al	Cetraria nivalis C. cuculata C. islandica C. richardsonii Thamnolia vermicularis Dactylina arctica Ecamadophila erictorum
	Exposed Alpine Lichens	l-ea	Alectoria spp. Stereocaulon spp.
	Rock Lichens	l-rk	Parmelia spp. Lecanora spp. Rhizocarpon spp.
	Arboreal Lichens	l-ar	Bryoria spp. Umbilicaria spp.

Table 11 Groups for Describing the Four Lower Strata within Major Vegetation Community of Biotic Land Unit

ORGANIC LITTER	Deadfall	o-df	dead woody material lying on ground; includes logs, large branches, and limbpiles
	Snags	o-sg	standing dead trees from which the leaves and most of the limbs have fallen
	Leaf Litter	o-ll	dead deciduous leaves and small twigs
	Needles	o-nd	dead conifer needles
	Cones	o-cn	conifer cones
SURFACE SUBSTRATE	Bedrock	s-bd	exposed consolidated mineral material
(all but the last of these groups may be covered by scattered mosses, lichens, ferns, or fern allies, or by organic or mineral layer less than 2 cm in thickness)	Cobbles and Stones	s-cs	exposed unconsolidated rock fragments greater than 7.5 cm in diameter
	Gravel	s-gr	exposed unconsolidated rock fragments less than 7.5 cm in diameter
	Sand	s-sd	mass of fine particles of crushed or worn rocks
	Mineral Soil	s-ms	unconsolidated mineral material of variable texture often associated with tree tip-ups, active erosion or deposition, severe fires, nivation areas, or exposed lake bottoms
	Peaty Soil	s-pt	exposed organic material ranging from decomposing vegetation parts to humified organic material
	Water	s-wt	small, regularly occurring areas of open water

[a] *After Oswald, 1980*
[b] *After Thomas, 1979*
[c] *After Walmsley, et al., 1980*

APPENDIX I

WILDLIFE ORIENTATION TO SIGNIFICANT HABITAT FEATURES - MAMMALS

MAMMALS	PREFERRED BREEDING HABITAT[a]	MAJOR FOOD SOURCE[b]	COVER REQUIREMENTS	ECOTONE	SELECTED REFERENCES
1. PIKA	- talus, cliffs, boulderfields beyond treeline to vegetation limit	- grasses, sedges, alpine forbs interspersed with or in close proximity to breeding habitat			(Banfield, 1974)
2. SNOWSHOE HARE	- "forms" in grass amongst rocks, stumps, deadfall in open forests and lowland shrubby area	- grasses and forbs particularly: calspp, brospp, oxyspp, astrspp, antspp, equspp; also poptre, salspp, betspp	- low shrubs	- interspersion of cover and forage areas	(CWS[d], 1972, Banfield, 1974)
3. HOARY MARMOT	- talus, rock slides, loose shale on open hummocky sites beyond treeline	- variety of grasses and alpine forbs	- large boulders		(Banfield, 1974)
4. ARCTIC GROUND SQUIRREL	- sandy to gravelly soil near forest clearings and beyond treeline, eg. eskers, moraines, river banks, dunes	- variety of grasses, forbs, low and ground shrubs			(Banfield, 1974)
5. RED SQUIRREL	- tree cavities, abundant cone and needle litter in conifer forest	- conifer cones, various forbs, fruits of ground shrubs; buds, flowers, fruit of poptre and betspp			(Banfield, 1974)
6. BEAVER	- low-gradient headwater or lowland streams flowing thrugh extensive organic overburden, ponds with abrupt steep-sided shoreline and island mats of vegetation; recently formed oxbow lakes	- riparian stands or associations of poptre; - riparian salspp in upland situations			(CWS, 1978a, Banfield, 1974)
7. PORCUPINE	- mixed or conifer forest with rocky ledges, boulders, or deadfall suitable for denning	- leaves of poptre or betpop; cambium of picspp or abilas			(CWS, 1977b, Banfield, 1974)
8. COYOTE	- prefers open habitats to treeline	- ground squirrels, snowshoe hare			(CWS, 1977c, Banfield, 1974)
9. WOLF	- known to den in all but the highest and wettest habitats	- ground squirrels, snowshoe hare, beaver, moose, caribou sheep			(CWS, 1973a, Banfield, 1974)
10. RED FOX	- known to den in all but the highest and wettest habitats	- ground squirrels, snowshoe hare, small mammals			(CWS, 1978b, Banfield, 1974)
11. BLACK BEAR	- mixed or conifer forest, or dense shrublands with deadfall boulders or caves suitable for denning	- berry-producing shrubs; shecan, empnig, arcuva, ribspp, vacspp, rubspp; calspp, carspp; roots of hedalp			(Banfield, 1974, Pearson, 1975, Smith, 1980 for both bear species)
12. GRIZZLY BEAR	- shrubby or semi-open areas above treeline with 30 to 40 degree slope, 1100-1300 metre elevation and non-northern aspect	- same as Black Bear		- forested areas adjacent to avalanche scars; timberline	
13. CANADA LYNX	- mature conifer forest with dense low or medium shrub under-story and rocky ledges or deadfall for bedding cover	- snowshoe hare			(CWS, 1977d, Banfield, 1974)
14. WOODLAND CARIBOU[c]	- calving areas too variable for inference purposes	- fruticose and arboreal lichens; salspp, carspp	- escape terrain		(CWS, 1973b, Banfield, 1974, Larsen, 1981)
15. MOOSE[c]	- calving areas too variable for inference purposes	- medium to tall willow shrubs; popspp, betspp	- lowland forests for thermal cover		(CWS, 1977e, Banfield, 1974)
16. MOUNTAIN GOAT[c]	- rugged alpine terrain with steep cliffs and vegetated ledges	- grasses, sedges, rushes and alpine forbs; stunted conifers above treeline	- caves for thermal cover; escape terrain		(Hoefs, Lortie, & Russel, 1977, Banfield, 1974)
17. DALL SHEEP[c]	- lambing areas too variable for inference purposes	- grasses, especially fesalt plus sedges and alpine forbs	- caves for thermal cover; escape terrain		(CWS, 1973c, Banfield, 1974, Geist, 1974, Murie, 1971)

[a] *Habitat in which young are born and raised*

[b] *Forage plants listed (alphabetical order): abilas Abies lasiocarpa, antspp Antennaria spp; arcuva Arctostaphylos uva-ursi, astspp Aster spp., betspp Betula spp., brospp Bromaus spp., calspp Calamagrostis spp., carspp Carex spp., empnig Empetrum nigrum, equspp Equisetum spp., hedalp Hedysarum alpinum, picspp Picea spp., poptre Populus tremuloides, ribspp Ribes spp., rubspp Rubus spp., salspp Salix spp., shecan Shepherdia canadensis, vacspp Vaccinium spp.*

[c] *Salt licks important to these species for the replenishment of nutrients following winter*

[d] *Canadian Wildlife Service*

Table 12 Significant Habitat Features for Selected Mammals Found in Sub-Arctic Yukon Territory, Canada

APPENDIX J

WILDLIFE ORIENTATION TO SIGNIFICANT HABITAT FEATURES - BIRDS

BIRDS	PREFERRED BREEDING AND/OR FORAGING HABITAT
1. Common Loon	- small lowland lakes bordered by tamarack or spruce
2. Horned Grebe	- small lowland lakes, ponds, marshes, sluggish rivers with graminoid margins and island mats
3. Ducks	- shallow creeks, ponds, marshes or protected bays with extremely irregular shoreline and interspersion of submergent and emergent vegetation with island mats
4. Goshawk	- lowland conifer or mixed forests
5. Sharp-Shinned Hawk	- closed conifer forest interspersed with shrublands; open conifer forests
6. Red-Tailed Hawk	- open forests or woodlands; open shrublands or graminoid meadows in vicinity of trees
7. Bald Eagle	- mixed or conifer shorelines of large lowland rivers or lakes
8. Golden Eagle	- nests on cliffs, canyons, jagged rock outcrops in mountains; forages over alpine or subalpine shrublands and bryoid mat
9. Northern Harrier	- nests on ground in low shrubs usually in hygric meadow, shrubland or wetland edge. forages over similar habitat
10. Osprey	- mixed or conifer shorelines or large lowland rivers or lakes
11. Gyrflacon	- nests on cliffs, canyons, jagged rock outcrops in mountains; forages over alpine or subalpine shrublands and bryoid mat
12. American Kestrel	- open country with scattered tall shrubs, snags, or trees, or in vicinity of cliffs
13. Peregrine Falcon	- nests on cliffs, canyons, jagged rock outcrops in mountains; forages over alpine or subalpine shrublands and bryoid mat
14. Blue Grouse	- burned over areas with tall shrubs; open subalpine conifer forests; conifer woodlands
15. Spruce Grouse	- conifer forests and woodlands; subalpine areas with scattered tall shrubs
16. Ruffed Grouse	- mature deciduous or mixed forests; burned over areas with tall shrubs
17. Willow Ptarmigan	- subalpine conifer woodlands; gently sloping medium shrublands over treeline
18. Rock Ptarmigan	- low or ground shrublands in apline areas; bryoid mat
19. White-Tailed Ptarmigan	- rocky alpine meadows, boulder fields, rock slides, and talus to vegetation limit
20. Semi-Palmated Plover	- level, sparsely-vegetated, sandy or gravelly shorelines of ponds, lakes or rivers
21. Common Snipe	- wetlands, hygric meadows, streamsides with low shrubs and areas of soft mud
22. Spotted Sandpiper	- level, muddy, sandy or pebbly shorelines of ponds, lakes, rivers or streams
23. Least Sandpiper	- wetlands with a high proportion of graminoid and moss cover
24. Wandering Tattler	- gravel bars of streams in subalpine and woodlands and shrublands
25. Lesser Yellowlegs	- hygric woodlands, wetlands, or burned over areas with tall shrubs
26. Northern Phalarope	- shallow, quiet ponds with graminoid and low shrub margins
27. Herring Gull	- rocky islands or boulder-strewn lowland lakeshores
28. Mew Gull	- marshy areas
29. Bonaparte's Gull	- conifer woodlands or open conifer forests in vicinity of small lakes
30. Arctic Tern	- sand and gravel beaches, sandspits, dunes near lakes
31. Great Horned Owl	- all types of lowland forests
32. Hawk Owl	- coniferous or mixed open forests or woodlands, burned over areas with snags
33. Short-Eared Owl	- hygric meadows, shrublands, or wetlands with low shrubs in lowland and subalpine areas
34. Nighthawk	- conifer woodlands or open medium or tall shrublands with patches of bare soil, sand or gravel
35. Belted Kingfisher	- unvegetated banks and earth cliffs along clear-running rivers and lakeshores
36. Common Flicker	- woodlands; open forests; burned over areas, with snags
37. Hairy Woodpecker	- forests of all types; shrublands with tall shrub component
38. Northern Three-Toed Woodpecker	- mature conifer forests; burned over areas with snags
39. Say's Phoebe	- mesic to xeric graminoid meadow with scattered low and medium shrubs; burned over areas with snags
40. Traill's Flycatcher	- shrubby margins of shallow ponds; shrub bogs; shrublands with tall shrub component
41. Least Flycatcher	- open deciduous or mixed lowland forests; shrublands in lowland areas
42. Western Wood Pewee	- open lowland forests of all types; shrublands in lowland areas
43. Olive-Sided Flycatcher	- burned over areas with tall shrubs abd snags; conifer woodlands; open conifer forests
44. Violet-Green Swallow	- nests on cliffs, canyons, jagged rock outcrops on mountain face; forages over shrublands with tall shrub component and snags

Table 13 Significant Habitat Features for Selected Birds Found in Sub-Arctic Yukon Territory, Canada

45.	Tree Swallow	- nests on flooded shores or wetlands with numerous standing dead trees near water; forages over lakes, large streams, ponds and marshes
46.	Bank Swallow	- nests in steep banks of sand, sandstone, clay and gravel along river embankments, road and rail cuts and lakeshores; forages over water and open land
47.	Barn Swallow	- nests on lowland cliffs, caves, buildings, and bridges in vicinity of open water; forages over water and open land
48.	Cliff Swallow	- nests on lowland cliffs, caves, buildings and bridges in vicinity of open water; forages over water and open land
49.	Gray Jay	- forests and woodlands of all types; tree bogs
50.	Black-Billed Magpie	- medium subalpine shrubland with scattered conifers; open forests with high deciduous component
51.	Horned Lark	- level to moderately sloping apline terrain with a minimum of vegetation cover; graminoid meadows; sparsely vegetated sand dunes
52.	Boreal Chickadee	- open or closed conifer or mixed forests; conifer woodlands
53.	Red-Breasted Nuthatch	- open or closed conifer or mixed forests; conifer woodlands
54.	American Robin	- open forests; medium shrublands with scattered conifers; tree and shrub bogs; shrublands with tall shrub component
55.	Varied Thrush	- closed, mesic, conifer or mixed forests
56.	Hermit Thrush	- open mixed or conifer forests; shrublands with tall shrub component
57.	Swainson's Thrush	- open conifer forest interspersed with tall shrubs to treeline
58.	Gray-Cheeked Thrush	- closed conifer forest with low or medium shrub understory especially near treeline
59.	Mountain Bluebird	- burned over lowland areas with tall shrubs and snags
60.	Townsend's Solitare	- conifer woodlands; open conifer and mixed forests
61.	Ruby-Crowned Kinglet	- closed conifer forests
62.	American Pipit	- alpine and subalpine shrublands; bryoid mat
63.	Bohemian Waxwing	- open or closed conifer or mixed forests; conifer woodlands; burned over areas with tall shrubs
64.	Tenessee Warbler	- deciduous or mixed forests; hygric or mesic, medium shrublands; shrub bog
65.	Orange-Crowned Warbler	- deciduous or mixed forests; burned over areas with medium shrubs
66.	Yellow Warbler	- shrubby margins of rivers and wetlands
67.	Yellow-Rumped Warbler	- forests of all types to treeline
68.	Balckpoll Warbler	- coniferous and mixed forests; tall shrublands with scattering of low conifers
69.	Common Yellowthroat	- hygric or mesic mixed forests with medium shrub understory; shrub bogs; marshes
70.	Wilson's Warbler	- medium and tall mesic shrublands from valley floor to upper slopes
71.	Rust Blackbird	- wetlands; hygric shrublands in lowland and subalpine areas
72.	Gray-Crowned Rosy Finch	- cliffs, canyons, and jagged rock outcrops on mountain face
73.	Common Redpoll	- subalpine shrublands with scattered conifers
74.	Pine Siskin	- coniferous or mixed forests and woodlands
75.	White-Winged Crossbill	- coniferous or mixed forests and woodlands
76.	Savannah Sparrow	- hygric meadows; wetlands; partially stabilized sand dunes with grass cover; subalpine shrublands; woodlands with shrub understory
77.	Dark-Eyed Junco	- forests and woodlands of all types to treeline
78.	Tree Sparrow	- alpine or subalpine shrublands; subalpine woodlands with shrub understory
79.	Brewer's Sparrow	- low and medium shrublands above treeline
80.	Chipping Sparrow	- conifer forests interspersed with medium or tall shrubs
81.	White-Crowned Sparrow	- low or medium shrublands from lowland to alpine areas; shrubby wetlands
82.	Golden-Crowned Sparrow	- low or medium shrublands at or above treeline
83.	Fox Sparrow	- medium mesic to hygric woodlands or shrublands with scattered conifers, especially near treeline; shrubby wetlands

Selected References: Erskine, 1977; Godfrey, 1976; Hoefs, 1976; Murie, 1963; Peterson, 1969; Theberge, 1974, 1976

University of Waterloo
Department of Geography Publication Series

Available from Publications, Department of Geography, University of Waterloo, Waterloo, Ontario, N2L 3Gl

23 BRYANT, Christopher, R., 1984, *Waterloo Lectures in Geography, Vol. 1, Regional Economic Development,* 114 pp.
22 KNAPPER, Christopher, GERTLER, Leonard, and WALL, Geoffrey, 1983, *Energy, Recreation and the Urban Field,* 89 pp.
21 DUDYCHA, Douglas J., SMITH, Stephen L.J., STEWART, Terry O., and McPHERSON, Barry D., 1983, *The Canadian Atlas of Recreation and Exercise,* 61 pp.
20 MITCHELL, Bruce, and GARDNER, James S., 1983, *River Basin Management: Canadian Experiences,* 443 pp.
19 GARDNER, James S., SMITH, Daniel J., and DESLOGES, Joseph R., 1983, *The Dynamic Geomorphology of the Mt. Rae Area: A High Mountain Region in Southwestern Alberta,* 237 pp.
18 BRYANT, Christopher R., 1982, *The Rural Real Estate Market: Geographical Patterns of Structure and Change in an Urban Fringe Environment,* 153 pp.
17 WALL, Geoffrey, and KNAPPER, Christopher, 1981, *Tutankhamun in Toronto,* 113 pp.
16 WALKER, David F., editor, 1980, *The Human Dimension in Industrial Development,* 124 pp.
15 PRESTON, Richard E., and RUSSWURM, Lorne H., editors, 1980, *Essays on Canadian Urban Process and Form II,* 505 pp. (No longer available).
14 WALL, Geoffrey, editor, 1979, *Recreational Land Use in Southern Ontario,* 376 pp.
13 MITCHELL, Bruce, GARDNER, James S., COOK, Robert, and VEALE, Barbara, 1978, *Physical Adjustments and Institutional Arrangements for the Urban Flood Hazard: Grand River Watershed,* 142 pp.
12 NELSON, J. Gordon., NEEDHAM, Roger D., and MANN, Donald, editors, 1978, *International Experience with National Parks and Related Reserves,* 624 pp.
11 WALL, Geoffrey, and WRIGHT, Cynthia, 1977, *Environmental Impact of Outdoor Recreation,* 69 pp.
10 RUSSWURM, Lorne H., PRESTON, Richard E., and MARTIN, Larry R.G., 1977, *Essays on Canadian Urban Process and Form,* 377 pp. (No longer available).
9 HYMA, B., and RAMESH, A., 1977, *Cholera and Malaria Incidence in Tamil, Nadu, India: Case Studies in Medical Geography,* 322 pp. (No longer available).
8 WALKER, David F., editor, 1977, *Industrial Services,* 107 pp.

7 BOYER, Jeanette C., 1977, *Human Response to Frost Hazards in the Orchard Industry, Okanagan Valley, British Columbia,* 207 pp. (No longer available).

6 BULLOCK, Ronald A., 1975, *Ndeiya, Kikuyu Frontier: The Kenya Land Problem in Microcosm,* 144 pp.

5 MITCHELL, Bruce, editor, 1975, *Institutional Arrangements for Water Management: Canadian Experiences,* 825 pp. (No longer available).

4 PATRICK, Richard A., 1975, *Political Geography and the Cyprus Conflict: 1963-1971,* 488 pp. (No longer available).

3 WALKER, David F., and BATER, James H., editors, 1974, *Industrial Development in Southern Ontario: Selected Essays,* 306 pp. (No longer available).

2 PRESTON, Richard E., editor, 1973, *Applied Geography and the Human Environment: Proceedings of the Fifth International Meeting, Commission on Applied Geography, International Geographical Union,* 397 pp.

1 McLELLAN, Alexander G., editor, 1971, *The Waterloo County Area: Selected Geographical Essays,* 316 pp.